T0235461

SpringerBriefs in Food, Health, and Nutrition

Springer Briefs in Food, Health, and Nutrition present concise summaries of cutting edge research and practical applications across a wide range of topics related to the field of food science, including its impact and relationship to health and nutrition. Subjects include:

- Food chemistry, including analytical methods; ingredient functionality; physic-chemical aspects; thermodynamics
- Food microbiology, including food safety; fermentation; foodborne pathogens; detection methods
- Food process engineering, including unit operations; mass transfer; heating, chilling and freezing; thermal and non-thermal processing, new technologies
- Food physics, including material science; rheology, chewing/mastication
- Food policy
- And applications to:
  - Sensory science
  - Packaging
  - Food quality
  - Product development

We are especially interested in how these areas impact or are related to health and nutrition.

Featuring compact volumes of 50 to 125 pages, the series covers a range of content from professional to academic. Typical topics might include:

- A timely report of state-of-the art analytical techniques
- A bridge between new research results, as published in journal articles, and a contextual literature review
- A snapshot of a hot or emerging topic
- An in-depth case study
- A presentation of core concepts that students must understand in order to make independent contributions

More information about this series at http://www.springer.com/series/10203

Anthony Keith Thompson

Fruit and Vegetable Storage

Hypobaric, Hyperbaric
and Controlled Atmosphere

 Springer

Anthony Keith Thompson
Huddersfield
UK

ISSN 2197-571X   ISSN 2197-5728   (electronic)
SpringerBriefs in Food, Health, and Nutrition
ISBN 978-3-319-23590-5   ISBN 978-3-319-23591-2   (eBook)
DOI 10.1007/978-3-319-23591-2

Library of Congress Control Number: 2015948785

Springer Cham Heidelberg New York Dordrecht London

Printed on acid-free paper

Springer International Publishing AG Switzerland is part of Springer Science+Business Media (www.springer.com)

Preface

The purpose of this book is to evaluate the effects of changing the pressure within the storage environment of fresh fruit and vegetables in order to preserve them in their optimum condition. This evaluation is put into context of current commercial practices used in *Fruit and Vegetable Storage* technology in order to determine the possible future of changing the pressure as a commercial technique. Its primary purpose is therefore to provide an up-to-date consideration of these techniques that will help those involved in the postharvest fruit and vegetable industry to plan and design the most suitable and economic conditions both now and in the future. Students in tertiary education will also find it useful both in giving an overview of the subject and to determine possible areas of research needs.

Thanks are due to Dr. Pansa Liplap for helpful comments on the manuscript and permission to use figures and Ken Hatch of UK CA Ltd. for useful comments.

Huddersfield, UK   Anthony Keith Thompson

Contents

Abbreviations

1-MCP	1-methylcyclopropene
ACC	1-aminocyclopropane-1-carboxylic acid
APX	Ascorbate peroxidase
Atm	Atmosphere
BCE	Before Common Era
CA	Controlled atmosphere storage
CAT	Catalase
CF	Chlorophyll fluorescence
DCA	Dynamic controlled atmosphere
DCA-CF	Dynamic controlled atmosphere-chlorophyll fluorescence
DPPH	(1,1-diphenyl-2-picrylhydrazyl)
EU	European Union
GM	Genetically modified
kPa	Kilo Pascals
MDA	Malondialdehyde
mm Hg	mm of mercury
MPa	Mega Pascals
N	Newton
nor	Non-ripening tomatoes
Pa	Pascal
PAL	Phenylalanine ammonia lyase
PG	Polygalacturonase
PME	Pectin methylesterase
POD	Peroxidase
r.h.	Relative humidity
rin	Ripening inhibitor tomatoes
RQ	Respiratory quotient
SAM	S-adenosyl-L-methionine
SI	Système International d'unités
SOD	Superoxide dismutase

SSC	Soluble solids content
TA	Total acidity or titratable acidity
TSS	Total soluble solids
VPD	Vapour pressure deficit

Introduction

Fruit and vegetables are crucial parts of the human diet but their condition can change after they have been harvested. These changes include chemical content, physical structure and those as a result of microorganism infections. In addition to these changes, there is also progressively increasing demand for perceived high quality fruit and vegetables, constant availability and maintenance of their nutritional and health benefits. Therefore the way they are stored during the postharvest period, be it simply during short marketing procedures or long term storage to link up seasonal availability, the environmental conditions in which they are kept can have a vital influence. Considerable research has been undertaken on the effects of postharvest environmental conditions on these changes in fruit and vegetables. Research has concentrated largely on the effects of temperature, humidity and environmental gases; mainly oxygen, carbon dioxide and ethylene. The manipulation of these environmental conditions has become standard commercial practice, but less consideration has been made of environmental pressure. However, changing the pressure around fruit and vegetables has been the subject of research over recent decades and some commercial application has been attempted but has been largely unsuccessful. The present book reviews the effects of both increasing and reducing the atmospheric pressure on the changes in the postharvest life of fruit and vegetables. It puts the studies in context of more common methods used in their preservation and describes the technology that has been used as well as evaluates the history and prospects of the use of hypobaric and hyperbaric storage. It concludes that both these techniques of changing the atmospheric pressure have potential for application to address quantitative and qualitative challenges in the postharvest sector of the fresh fruit and vegetables industry. There are reports of the effects of hypobaric storage on at least 45 fruit and vegetables as well as on whole plants and cut flowers while only eight research reports into hyperbaric storage of fruit and vegetables could be found.

Chapter 1
Storage

Introduction

The technology involved in getting fresh produce from the field to the consumer has been the subject of detailed scientific research for well over a century. Although fruit and vegetables are mainly seasonally available; as time has progressed and living standards have improved many consumers have developed a requirement for specific fresh fruit and vegetables to be available whenever they want them. This requirement has been stimulated by the retail trade and the technological improvements of producing out-of-season crops as well as the improvement in international transport. However, perhaps the main contribution to their prolonged availability has been the development of techniques in the storage of fruit and vegetables after harvest. This has been mainly the development of refrigeration in the nineteenth century but alongside this controlled atmosphere storage and modified atmosphere packaging were developed in the twentieth century and used to supplement refrigeration. Changing the moisture content of the air in the refrigerated store has also been shown to have an effect on the retention of quality and the rate of deterioration of many fruit and vegetables and together with temperature and gaseous atmosphere within the store has been developed to highly sophisticated levels. So successful have these developments been that perhaps many people do not even realise that some fruit and vegetables used to have only seasonal availability.

Many fresh fruits and vegetables deteriorate quickly after they have been harvested. The rate of deterioration depends on many factors including the way they are stored. Storage conditions may extend their marketability from only a few days (or even a few hours) to several months. Their postharvest behaviour also depends on type; for example fruits can be classified into two groups, climacteric and non-climacteric that have very different harvest and postharvest requirements. The term climacteric was first applied to fruit ripening by Kidd and West (1927).

© The Author(s) 2016
A.K. Thompson, *Fruit and Vegetable Storage*, SpringerBriefs in Food,
Health, and Nutrition, DOI 10.1007/978-3-319-23591-2_1

Climacteric fruit, such as bananas and tomatoes, can be ripened after harvest and non-climacteric, such as oranges, do not ripen after harvest. Non-climacteric fruit should therefore be harvested only when fully mature. Climacteric fruit go through a rise in respiration rate, which usually coincides with fruit softening, flavour development and changes in colour. Ripening is initiated when the level of ethylene reaches a critical level in the fruit cells. In non-climacteric fruit, e.g. strawberry there was little change in ethylene production during ripening (Manning 1993). Ripening of climacteric fruit can also be initiated by exposing the fruit to exogenous ethylene after harvest so that the threshold level infiltrates into the cells of the fruit. As climacteric fruits approach maturity they become more sensitive to exogenous ethylene (Knee 1976). Ethylene can also have negative effects on fresh fruit and vegetables. For example, Wills et al. (1999) showed that the postharvest life of some non-climacteric fruit and vegetables could be extended by up to 60 % when stored in less than 5 parts per billion ethylene compared with those stored in 100 parts per billion ethylene and Rees (2012) reported that in Britain, ethylene can accumulate in packhouses to concentrations near 1000 parts per billion. This will be discussed later.

History

The major way of prolonging the postharvest life of fresh fruit and vegetables is refrigeration. Koelet (1992) reported that even as early as 2000 BCE there is evidence that inhabitants of Crete were aware of the importance of temperature in the preservation of food. Toole (1990) reported that the practical use of reduced temperatures to preserve food dates back at least to 1750 when ice houses were first used. In Europe ice houses were lined pits below ground where ice was stored that had been taken from frozen lakes during winter. Mechanical refrigeration was first developed in 1755 by William Cullen who showed that evaporating ether under reduced pressure, caused by evacuation, resulted in the temperature of the water in the same vessel to be reduced and form ice. Cullen then patented a machine for refrigerating air by the evaporation of water in a vacuum. John Leslie subsequently developed Cullen's technique in 1809 by adding sulphuric acid to absorb the water vapour. In 1834 Jacob Perkins was granted a British patent (Serial Number 6662) for a vapour compression refrigeration machine. Carle Linde developed an ammonia compression machine that was used in the first sea freight shipment of chilled meat from Argentina in 1879 and the design was employed in many of the early experiments in refrigerated fruits and vegetable stores. The type of refrigeration used was mostly brine-circulated secondary systems. A simple refrigeration unit consists of an evaporator, a compressor, a condensed and an expansion valve. The evaporator is the pipe that contains the refrigerant mostly as a liquid at low temperature and low pressure. This pipe passes inside the store. The evaporator causes the refrigerant to evaporate and absorb heat. This vapour is drawn along the pipe through the compressor, which is a pump that compresses the gas into a hot high-pressure vapour. This is pumped to the condenser where the gas is cooled by passing it through a radiator. The radiator is usually a network of pipes open to the atmosphere. The high-pressure liquid is passed through a series of small bore pipes that slows down the flow of liquid so that the high pressure builds up. The liquid then passes through an expansion valve that controls the flow of refrigerant and reduces its pressure. This reduction in pressure results in a reduction in temperature causing some of the refrigerant to vaporise. This cooled mixture of vapour and liquid refrigerant passes into the evaporator so completing the refrigeration cycle. In most stores a fan passes the store air over the coiled pipes containing the refrigerant that helps to cool the air quickly and distribute it evenly throughout the store. Refrigeration systems are categorised by the method that the refrigerant is fed to the evaporator as direct expansion, natural flooded or pump circulation.

Refrigeration has also been developed for transport of fresh fruit and vegetables. A refrigerated rail car was patented in the USA in 1867 with ice bunkers in each end. Air was passed from the top through the cargo. The air flow was controlled by hanging flaps and was circulated by gravity. The first refrigerated car reported to carry fresh fruit in the USA was in 1867. It contained 200 quarts of strawberries together with 100 pounds of ice.

In modern times, since the first application of mechanical refrigeration to food transport by sea freight in 1879, the international trade in fresh fruit and vegetables largely uses temperature controlled cargo space in either "break bulk" in the refrigerated holds of ships or in refrigerated containers called "reefers". International sea transport of fruit and vegetables and other perishable foodstuffs, using refrigeration, began in the latter part of the nineteenth century as break bulk. Reefer containers for transport by sea were first introduced in the 1930s but it was only in the 1950s that large numbers of reefer containers were transported on ships. The first purpose built container ship was completed in 1969 which had a capacity of 1,300 twenty-foot equivalent container units. Reefer containers are also transported on land on purpose built trucks often between packhouses and ports and from ports to distribution centres or packhouses.

Changes During Storage

The postharvest changes in fresh fruit and vegetables are affected by their postharvest environment as well as micro-organism infection, the stage of their development or maturity at harvest and the conditions in which they have been grown. The changes also depend on the part of the plant or tree on which it has grown. Some vegetables are natural storage organs, for example potato tubers and onions bulbs, and their postharvest changes are different from say leaf vegetables, such as lettuce and cabbages. The postharvest requirements for fruit also can vary considerably. Fruits are often classified into climacteric and non-climacteric and, in some cases, intermediate where their ripening metabolism is not clear. Climacteric fruits are those whose ripening is accompanied by an increase in respiration rate, called the climacteric rise, which is generally associated with increased ethylene production. Initiation of the climacteric rise in respiration rate is by ethylene biosynthesis and associated with other chemical and physical changes. There are more postharvest changes in climacteric fruit than in non-climacteric fruit or in vegetable. In non-climacteric fruit or vegetable the chemical content remains similar during their postharvest life except for perhaps sugars which are utilised for metabolic processes and therefore decrease. In climacteric fruit there are considerable changes that we commonly refer to as ripening where the fruit develops typical flavour and aroma, changes colour (through loss of chlorophyll and synthesis of carotenoids and other pigments), changes of starch into sugars and changes in the cell wall constituents that probably contribute to softening. Cell walls are complex structures composed of cellulose and pectin, derived from hexoses including glucose, galactose, rhamnose and mannose, as well as pentoses including xylose and arabinose and some of their derivatives including glucuronic acid and galacturonic acid. The changes in aroma volatile chemicals are important since they affect the acceptability of fruit and vegetables.

Among other factors the levels and type of phytochemicals in fruit and vegetables depend on temperature, humidity and gaseous environment. Carbohydrates are lost rapidly when the fruit or vegetables are subject to conditions which increase their respiration rate. Other postharvest nutritional losses were shown by Alighourchi et al. (2008) who found that the mean degradation percentage of the 15 pomegranate anthocyanins studied was between 23 and 83 % during 10 days at 4 °C. Antioxidant activity increased during storage of four Italian apple cultivars, which correlated with an increase of the concentration of catechin and phloridzin (Napolitano et al. 2004). The flesh of the apple cultivar Annurca had antioxidant properties comparable to those in the peel and polyphenolic compounds are relatively stable in the peel and the flesh; which should be maintained during long-term storage (D'Angelo et al. 2007).

Experiments on lettuce stored for 15 days showed that storage at 10 °C affected their leaf quality with significant chlorophyll reduction after only 5 days. Total carotenoids significantly decreased after 8 days, but there were no significant changes in anthocyanins and total phenols over the 15-day storage period (Ferrante and Maggiore 2007). Fahey et al. (1997) demonstrated that 3-day-old broccoli sprouts contained 10 to 100 times higher levels of glucoraphanin than did corresponding mature plants. Sugar levels of carambola remained constant during storage, although fruits continued to lose chlorophyll and synthesise carotenoids (Wan and Lam 1984). Perkins-Veazie et al. (2008) found that total anthocyanin, total phenolics, and ferric reducing antioxidant power increased in blueberry fruit during storage at 5 °C for 7 days plus 2 days at 20 °C and 90 % r.h. Several cultivars of orange stored at 6 ± 1 °C for 65 days by Rapisarda et al. (2008) had reduced flavanone concentration, but increased vitamin C in blond cultivars. Antioxidant activity increased during storage caused mainly by phenolic accumulation in blood cultivars and vitamin C increase in blond cultivars. Stevens et al. (2008) found reduced ascorbic acid levels of tomato fruit following storage at 4 °C. In broccoli the antioxidant activity and L-ascorbic acid content increased during 4 weeks storage at 5 °C (Wold et al. 2007). Storage at 0 °C induced an accumulation of anthocyanin in the skin of grapes (Romero et al. 2008). Total carteniods in cherry tomatoes increased from 3297 µg 100 g^{-1} fresh weight in green fruit to 11694 µg 100 g^{-1} fresh weight in dark red fruit (Laval Martin et al. 1975). This colour development occurred in both the pulp and the flesh of the tomato. Rapisarda et al. (2008) evaluated eight citrus cultivars and found an increase in anthocyanins, flavanones and hydroxycinnamic acids and a slight decrease in vitamin C in the blood oranges during storage at 5–7 °C for 65 days. Cold storage negatively affected flavanone concentration, while it positively influenced vitamin C in blond orange cultivars. There was an increase in antioxidant capacity during cold storage caused mainly by phenolic accumulation in blood oranges and vitamin C increase in blond oranges.

Ripening involves a number of physical and chemical changes that occur in fruit. These normally occur when the fruit has developed to what is referred to as full maturity. However, immature fruit may be harvested and exposed to certain postharvest conditions (temperature, gas content in the atmosphere, humidity) that are conducive to ripening. The changes that can occur during ripening may be independent of each other. But under the correct circumstances these processes are initiated together and proceed together producing an acceptably ripe fruit. Probably the most obvious change in many fruits during ripening is their peel colour. For example, in tomatoes chlorophyll levels are progressively broken down into phytol. Seymour (1985) found that the chlorophyll content of banana peel progressively reduced during ripening but their carotenoid content could change depending on temperature. Those ripened at 35 °C had significantly increased carotenoids in the peel, while in those ripened at 20 °C carotenoids remained constant. During ripening of sapodillas at 22–28 °C total chlorophyll, tannins and total carotenoids decreased and softening increased, although the postharvest patterns differed with cultivar (Guadarrama et al. 2000). The carotenoid content in the aril of Bitter Gourd (*Momordica charantia*) increased during ripening, lycopene being the main component (Tan et al. 1999). Fruits normally soften progressively during ripening and although exact biochemical mechanisms have not yet been fully established, it is believed that softening is largely due to the breakdown of starch and other non-pectic polysaccharides and changes in the structure of cell wall components especially pectic polymers thereby reducing cellular rigidity. During the developmental stage of climacteric fruit there is a general increase in starch content, which is hydrolysed to simple sugars. For example, Medlicott and Thompson (1985) showed that the starch content of mangoes was completely hydrolysed to sugar during ripening. In bananas ripening involves a reduction in starch content from around 15–25 % to less than 5 % in the ripe pulp, coupled with a rise of similar magnitude in total sugars. During the early part of ripening sucrose is the predominant sugar, but in the later stage glucose and fructose predominate. Acidity of fruits generally decreases during ripening. Medlicott and Thompson (1985) showed that in mangoes the principal acids were citric and malic and that there was a large decrease in citric acid during ripening, but only a small decrease in malic acid. Desai and Deshpande (1975) found that the ascorbic acid content of bananas increased during ripening at 20 °C for 21 days and then decreased slightly to the end of the experiment (after 35 days). Wold et al. (2007) found that antioxidant activity and L-ascorbic acid increased during ripening of tomatoes, but there were no significant differences in the antioxidant activity between postharvest ripened fruit and those fruit that were allowed to ripen on the plant.

Factors Influencing Storage

As indicated above storage factors that can influence postharvest life of fruit and vegetables include: temperature, gaseous environment (especially O_2 CO_2 and ethylene), humidity and pressure. Pressure (hyperbaric and hypobaric) effects will be dealt with in subsequent chapters.

Temperature

Storage temperature can affect the chemical content, chemical reactions and physical integrity in fresh fruit and vegetables. These changes have been well documented in many publications, for example Thompson (2015). Lana et al. (2005) showed that lycopene in tomatoes increased during storage at 8 or 16 °C while at lower temperatures there was no change or a slight decrease. The optimum temperature for colour development in tomatoes is 24 °C; at 30 °C and above lycopene is not formed (Laval Martin et al. 1975). The mean ascorbic acid level was 18.8 mg 100 g^{-1} in pineapples fruit that had been stored at 10 °C for 21 days followed by 48 h at 28 ± 2 °C compared with 9.3 mg 100 g^{-1} in fruit stored at 28 ± 2 °C for the full 23 days (Wilson Wijeratnam et al. 2005). Also in pineapples Zhou et al. (2003) showed that an increase in polyphenoloxidase activity was related to the incidence of blackheart. Immature and over-mature fruits developed less blackheart than mature fruit and it was related to storage at 6–18 °C and did not occur at 25 °C. Ferrante and Maggiore (2007) found that storage at 10 °C affected lettuce leaf quality. Significant chlorophyll reduction was observed after only 5 days of storage. Total carotenoids significantly decreased after 8 days at both 4 and 10 °C. Anthocyanins and total phenols did not change significantly during the entire experimental period at either temperature. The level of total phenolics was stable in broccoli during storage for 7 days at 5 °C (Leja et al, 2001). Broccoli heads were stored at temperatures of 1 or 4 C and 99 % r.h. for up to 28 days but no deleterious effect on the levels of aliphatic glucosinolates and flavonols was found (Winkler et al. 2006). Du Pont et al. (2000) found up to 46 % loss of flavenoid glycoside content in lettuce and endive after 7 days storage at 1 °C. Perkins-Veazie (2008) found that total anthocyanin, total phenolics, and ferric reducing antioxidant power increased in blueberry fruit during storage at 5 °C for 7 days plus 2 days at 20 °C with 90 % r.h.

Storage conditions can affect the development of disorders related to temperature. Storage temperature can affect the rate of reaction and change in chemical processes within the fruit or vegetable and where the temperatures are too low or exposure to a particular temperature is too long; a physiological disorder can occur that is called chilling injury. For example, Ribeiro et al. (2005) reported that chilling injury occurred in arracacha (*Arracacia xanthorrhiza*) in storage at 5 °C. The symptoms of chilling injury were the development of irregular pit lesions on the whole tuber surface, followed by intense internal discoloration. Apples are normally stored at 0 °C but some apple cultivars suffer from chilling injury at temperatures in storage above 0 °C. For example, the apple cultivar Cox's Orange Pippin grown in Britain can suffer from chilling injury when stored below 3 °C. However, this effect is not a simple one and Cox's Orange Pippin grown in New Zealand can be successfully stored at 0 °C. Chilling injury symptoms include a brown discoloration of the cortex with streaks of darker brown in the vascular region and the tissue remaining moist (Wilkinson 1972). In cranberries chilling injury occurred in storage below 2.5 °C as a change in flesh texture from crisp to rubbery and a loss of natural lustre (Fidler 1963). At temperatures at, or below, about 12 °C green banana fruit develop a dull, often grey skin colour, starch is no longer converted to sugar and they subsequently fail to ripen properly and eventually become black and decay. To avoid this problem in international trade, bananas are shipped at 13–14 °C.

Storage temperature can affect the postharvest development of diseases due to micro-organism infection. There are many examples of this effect including Rhizopus Rot in apricots. This disease is caused by the fungus *Rhizopus stolonifer* and was reported to occur frequently in ripe or near-ripe fruit stored at 20–25 °C. Rapid cooling the apricots and storage below 5 °C was reported to be an effective control method (Crisosto and Kader 2002). Grey mould (*Botrytis cinerea*) on gooseberries develops rapidly at temperatures above 10 °C but storage at 0 °C prevents or slows disease development (Dennis 1983). Suzuki and Okabayashi (2001) found that the occurrence of lesions caused by *Alternaria alternata* in okra was severe during storage at 20 °C or higher, while they were only slight during storage at 10 °C. Yusuf and Okusanya (2008) found *Rhizopus stolonifer, P. oxalicum* and *Aspergillus niger* infecting yams (*Dioscorea rotundata*) during storage and that the optimum temperature for growth of these three fungi was the same at 35 °C, while rot development was inhibited on the tubers at 15 °C.

Humidity

Storage humidity mainly affects the rate of water loss from fruit and vegetables. Desiccation above a certain level can result in them being unacceptable to the market, but it can have other effects. For example, Thompson et al. (1974) showed that low humidity can also hasten ripening of plantains. This effect was confirmed by Ferris et al. (1993) who showed that fruits stored at 20 °C and 96–100 % r.h.

ripened about 15 days later than those stored at 20 °C and 55–65 % r.h. This effect may well have been due to the damage caused to the fruit by the rapid weight loss resulting in the initiation of wound ethylene production. Shin et al. (2008) found no differences between 65 and 95 % r.h. in colour change, flavonoid, phenolic concentrations, and total antioxidant activity during 12 days storage of strawberries. Somboonkaew and Terry (2011) reported that there were no significant effects of humidity, over the range of 80–100 % r.h, on the loss of sugars in litchi during 9 days storage at 13 °C but all three of the anthocyanins they measured decreased except in fruit stored at 100 % r.h. where they increased.

Oxygen

Fruit and vegetables, like most other organisms produce the energy required for maintaining their metabolism through aerobic respiration. Therefore, reduced availability of O_2 can reduce respiration rate and thus prolong their postharvest life. Jacques Etienne Berard at the University of Montpellier in 1819 (Berard 1821) found that harvested fruit absorbed O_2 and gave out CO_2 and that fruit stored in atmospheres containing no O_2 did not ripen, but if they were held for only a short period and then placed in air they continued to ripen. Kidd and West (1925), Blackman (1928), Blackman and Parija (1928) also described the relationship between apple ripening, O_2 tension and respiratory activity. In 1929 a commercial store for apples was built by a grower near Canterbury in Britain where the level of both O_2 and CO_2 were controlled along with temperature to extend their storage life. Since then controlled atmosphere storage has become standard practice in many situations for apples and an increasing number of other fruit and some vegetables (Thompson 2010). Many chemical reactions in plant cells are oxidative therefore the rate of metabolism may be reduced especially at very low levels of O_2 in stores. However, if the O_2 is below a threshold the tricarboxylic acid cycle is inhibited but the glycolytic pathway may continue resulting in a build up of acetaldehyde and ethanol which are toxic to the cells if allowed to accumulate. In modern controlled atmosphere apple stores, concentrations down to 1 % O_2 are used with close control and warnings where anaerobic respiration (fermentation) may begin.

Postharvest diseases and disorders of fruit and vegetables can be affected by O_2 and CO_2 content in the atmosphere. There is strong evidence that controlled atmosphere storage can reduce some diseases of cabbages. Storage at −0.5 to 0 °C in 5–8 % CO_2 prevented the spread of *B. cinerea* and total storage losses were lower in controlled atmosphere storage than in air (Nuske and Muller 1984). Pendergrass and Isenberg (1974) also reported less disease in stored cabbages, also mainly caused by *B. cinerea*, and better head colour was observed with storage in 5 % CO_2 + 2.5 % O_2 + 92.5 % N_2 compared to those stored in air at 1 °C and 75 %, 85 % or 100 % r.h. Berard (1985) described experiments in which 25 cabbage cultivars were placed at 1 °C and 92 % r.h. He found that those in 2.5 %

O_2 + 5 % CO_2 usually had reduced or zero grey speck disease and reduced incidence and severity of vein streaking compared to those stored in air for up to 213 days but not in every case. Black midrib and necrotic spot were both absent at harvest but in comparison with storage in air those stored in 2.5 % O_2 + 5 % CO_2 had increased incidence of black midrib and it also favoured the development of inner head symptoms on susceptible cultivars. In controlled atmosphere storage the incidence of necrotic spot in the core of the heads of cultivar Quick Green Storage was increased which was particularly evident in a season when senescence of cabbage was most rapid. Even though both disorders were initiated in the parenchyma cells, black midrib and necrotic spot had a distinct histological evolution and affected different cultivars under similar conditions of growth and storage (Berard et al. 1986).

Storage of ginseng roots at low temperatures or controlled atmosphere storage slowed the rate of development and the rate of spread of grey mould (*B. cinerea*) (Lee and Yun 2002). Controlled atmosphere storage has been shown to reduce disease development in grapes (Massignan et al. 1999). Also it was shown that crown rot in bananas could be partially controlled by packing bananas in modified atmosphere packaging (Bastiaanse et al. 2010), which may reflect the gaseous environment within the bags.

Carbon Dioxide

The effect of CO_2 in extending the storage life of crops appears to be primarily on reduction of their respiration rate. Knee (1973) showed that CO_2 could inhibit an enzyme (succinate dehydrogenase) in the tricarboxylic acid cycle, which is part of the crop's respiratory pathway. An increased level of CO_2 is commonly used in controlled atmosphere stores and has been shown to have many other beneficial effects (Thompson 2010). Enhancement of CO_2 has also been used in hypobaric storage but this is uncommon (Burg 2004).

High CO_2 levels in storage have been generally shown to have a negative effect on the growth and development of disease-causing micro-organisms. There is also some evidence that fruit develop less disease on removal from high CO_2 storage than after previously being stored in air. However, in certain cases the levels of CO_2 necessary to give effective disease control have detrimental effects on the quality of the fruit or vegetable. The mechanism for reduction of diseases appears to be a reaction of the fruit rather than directly affecting the micro-organism, although there is some evidence for the latter. Exposing fruit and vegetables to high levels of CO_2 can also be used to control insects infecting fruit and vegetables, but again extended exposure to insecticidal levels of CO_2 may be phytotoxic. Exposure to high levels of CO_2 can affect physiological disorders for example Wang (1990) reviewed the effects of CO_2 on brown core of apples and concluded that it was due to exposure to high levels of CO_2 at low storage temperatures.

Ethylene

Ethylene is produced by plants usually in very low concentrations. However, in climacteric fruit, the biosynthesis of copious quantities of ethylene initiates their ripening. The biosynthesis of ethylene in ripening fruit was first shown by Gane (1934), who also found that ethylene biosynthesis stopped in the absence of O_2. Ethylene is synthesised in plants from methionine which is converted to S-adenosyl-L-methionine (SAM) catalysed by SAM synthetase, which is converted to 1-aminocyclopropane-1-carboxylic acid (ACC) catalysed by ACC synthase which is finally converted to ethylene catalysed by ACC oxidase. Two systems of ethylene production have been defined in climacteric fruit. System 1 functions during normal growth and development and during stress responses, whereas System 2 operates during floral senescence and fruit ripening. System 1 is autoinhibitory, such that exogenous ethylene inhibits synthesis and inhibitors of ethylene action can stimulate ethylene production. System 2 is stimulated by ethylene and is therefore autocatalytic and inhibits ethylene production and action (McMurchie et al. (1972). System 2 affects ACC synthase and ACC oxidase activity (Bufler 1986). The chlorophyll in the peel of Cavendish bananas normally breaks down during ripening revealing the yellow colour of the carotenoids pigments. Chlorophyll does not break down sufficiently at temperature over 25 °C and the fruit may remain green but ripen in all other respects. This effect is called *Pulpa crema* in Latin America and is where banana fruit are initiated to ripen on the plant because of stress due to insufficient water or fungal infections of the leaves particularly Leaf Spot and Sigatoka (*Mycosphaerella* spp.). This is presumably due to the biosynthesis of System 1 ethylene (Seymour 1985; Blackbourn et al. 1990).

Citrus, as non-climacteric fruit, evolve very low level of ethylene during natural maturation. A detailed analysis of the expression of ethylene biosynthetic and signalling genes during the whole reproductive development of Valencia orange fruit revealed a shift in ethylene regulation (Katz et al. 2004). In young fruitlets ethylene production is high and coincident with June drop, suggesting the hormone may be involved in the abscission process of developing fruits. At this stage, application of exogenous ethylene, or the analogous propylene, stimulated ethylene production in a feedback mechanism similar to that operative in climacteric fruits, that is classically refereed as System 2 (McMurchie et al. 1972). Interestingly, mature-coloured fruit producing very low ethylene responded to exogenous treatment in an autoinhibitory fashion, in agreement with previous data (Riov and Yang, 1982), and characteristic of the referred System 1. Together, these results indicate that ethylene biosynthesis during reproductive development of citrus fruits shifts from System 2 in developing fruits to the System 1 in mature fruits, a process that is opposite to that operative in climacteric fruits (Katz et al. 2004). The two systems can also occur in non-climacteric fruit such as oranges. Application of exogenous ethylene to young citrus fruits stimulated ethylene production in a feedback mechanism similar to System 2 in climacteric fruits. Mature citrus fruit produce very low ethylene responses to exogenous ethylene in an autoinhibitory fashion; System 1. These observations indicate that ethylene biosynthesis during reproductive development of citrus fruits shifts from System 2 in developing fruits to the System 1 in mature fruits, a process that is opposite to that in climacteric fruits (Katz et al. 2004). In responses to stress, such as wounding citrus fruit also show increased ethylene production. Under these conditions, the two members of the ACC synthase genes are regulated in a stress-specific manner and are not stimulated by ethylene.

Changing the atmospheric pressure within the fruit and vegetable store has also been shown to have effects on the postharvest life of many fruits and vegetables but for various reasons, which will be considered in this book, it has had limited commercial application. Most of the work on the effects of pressure on fresh fruit and vegetables has been with reduced pressure also called hypobaric pressure. Other work has been on increased atmospheric pressure also called hyperbaric pressure. Goyette (2010) defined hypobaric conditions as exposure to 0–0.1 MPa and hyperbaric conditions as exposure to 0.1–1 MPa applied to fresh horticultural crops and over 100 MPa to processed food.

Exposing harvested fruit and vegetables to ethylene can stimulate respiration rate, initiate climacteric fruit to ripen and can result in the rapid breakdown of chlorophyll (Thompson and Seymour 1982). This effect has been shown on a wide variety of crops including celery, cucumbers, cabbage, Brussels sprouts, cauliflower leaves, capsicums, tomatoes, broccoli and citrus fruits (Thompson 2015). Application of ethylene to stored tomatoes was shown to increase their carotenoid and lycopene content (Salunkhe and Wu 1973). This appears to be additional to the increase which is normally associated with ripening. Curd (1988) showed that strawberries exposed to ethylene had a more intense red colour than those stored in ethylene free air. In peas it can stimulate the formation of the toxin phytoalexin pisatin (Chalutz and Stahman 1969) and in sweetpotatoes the production of phenolics (Solomos and Biale 1975). Carrots exposed to ethylene can synthesise isocoumarin, which gives them a bitter flavour. Isocoumarin content was shown to increase with increasing concentrations of ethylene over the range of 0.5–50 µl L^{-1} (Lafuente et al. 1989). Sdiri et al. (2012) tested degreening with 2000 µl L^{-1} ethylene for 120 h at 21 °C with 95 % r.h. followed by quarantine treatment at 1 °C for 16 days then shelf life at 20 °C and 95 % r.h. for 7 days on early-season citrus fruits. They found some changes in individual flavonoid compounds following these conditions but these changes did not contribute to a loss in the total content of flavanones and flavones. Also they did not induce detrimental changes in DPPH (1,1-diphenyl-2-picrylhydrazyl) and ferric reducing antioxidant power, total ascorbic acid or total phenolic content.

Genetic Effects on Storage

With the rapid development in research on genetics and the publication of the complete genetic makeup of many individual plant and animal species it is important to consider how this effects the postharvest behaviour of fresh fruit and vegetables. Clearly, the genetic makeup affects their postharvest life and it has been known for perhaps hundreds of years that different varieties or cultivars of the same species behave and deteriorate differently during growth and after harvest. Recently specific genes have been shown to affect different postharvest changes. An example of this is the report by Shijie Yan et al. (2013) with storage of Yali pears (*Pyrus bretschneideri*). They reported that polyphenol oxidase gene expression increased and then decreased in core tissues during storage but rapid cooling promoted the gene expression. Hasperué et al. (2013) reported that in genes associated with chlorophyll degradation in broccoli, changes in the metabolism due to the time of day when they were harvested not only influence the expression of genes during the day but also may cause different patterns of expression during the postharvest period. Blauer et al. (2013) reported that in potato tubers ascorbate concentration began to fall during vine senescence and continued to decline progressively through maturation and storage, which was consistent with low levels of gene expression. Xingbin Xie et al. (2014) discussed challenges in ripening d'Anjou pears (*Pyrus communis*) after they had been treated with 1-methylcyclopropene to control superficial scald. In experiments with different storage and ripening conditions they concluded that a storage temperature of 1.1 °C can facilitate initiation of ripening capacity in 1-MCP treated pears with relatively low scald incidence following 6–8 months storage, through recovering the expression of certain ethylene synthesis and signal genes. Bo Zhang et al. (2014) suggested that reduced levels of volatiles associated with fruity aromas (such as esters and lactones) in the peach cultivar Hujingmilu (stored at 5 °C to cause chilling injury) were the consequence of modifications in expression of PpLOX1, PpLOX3 and PpAAT1 genes. Woolliness is a physiological disorder in peaches, which is characterised by a mealy texture, poor flavour and low juice content. Pavez et al. (2013) found that in the woolly fruit, upregulation of stress response genes was accompanied by downregulation of key components of metabolic pathways that were active during peach ripening. They suggested that the altered expression pattern of these genes might account for the abnormal ripening of woolly fruit. Harb et al. (2012) observed that internal ethylene concentrations in Honeycrisp apples were lower than in McIntosh, but Honeycrisp maintained their firmness while McIntosh softened rapidly during 10 days at 20 °C. At comparable internal ethylene concentrations, the expression of genes involved in ethylene synthesis, ethylene perception and signal transduction was generally much higher in Honeycrisp than in McIntosh. They concluded that this effect on softening did not appear to be related to expression of genes involved in ethylene biosynthesis. The temperature at which some fruit and vegetables suffer from chilling injury varies, not only with species, but also with cultivar (Thompson 2015). Aghdam (2013) reported that alternative oxidase was a candidate gene for manipulation of resistance to chilling injury. Postharvest treatments increased the alternative oxidase gene expression and can be used as a method for reducing their chilling injury. The practical application of gene expression for reducing chilling injury are based mainly on the application of signalling molecules such as salicylic acid, methyl jasmonate, methyl salicylate or jasmonic acids. Alternative oxidase gene expression is also enhanced by elevated O_2 atmospheres. Zheng et al. (2008) found that

Zucchini squash (*Cucurbita pepo*) were shown to develop chilling injury during storage at 5 °C, but the expressions of alternative oxidase were induced slightly when they were stored in 60 and 100 % O_2 for 3 days when compared with those stored in 21 % O_2. These increases in alternative oxidase transcript levels were correlated with the increased chilling resistance in the treated Zucchini squash.

Conventional plant breeding has been and continues to be used to improve desirable characteristics of plants we use for food. However, genetic modification has great potential for changing plant characteristics, including increasing the postharvest life, but especially disease control where genes are inserted into a crop that gives it resistance to a particular disease. However, there is resistance to genetic modification from politicians and the public of many countries. In September 2003 over 50 countries signed the Cartagena Protocol on biodiversity so that any country can refuse to import any genetic modified organism (USA was one of the countries that refused to sign the Protocol). However, genetic modified fruit and vegetables are allowed for consumption in many countries with no apparent detriment. For example, Solo papaya (*Carica papaya*) cultivars that are cultivated commercially in Hawai'i include: Rainbow, Sunup and Laie Gold that were all produced by genetic modification for resistance to the papaya ring spot virus. Genes for terpene syntases have been identified and characterised in *Citrus* spp. and their genetic modification was shown to alter their profile of volatile chemicals in the transgenic plants (Shimada et al. 2004). Rodríguez et al. (2011a) genetically modified *Citrus* spp. and generated Orange plants (*C. sinensis*) transformed with the limonene synthase gene taken from Satsumas (*C. reticulata*) in antisense orientation and generated mature fruits with a reduced content and emission of limonene. These trees produced fruit that showed substantial resistance to the pathogenic bacterium *Xanthomonas citri*, the pathogenic fungus *P. digitatum* and the Mediterranean fruit fly *Ceratitis capitata* (Rodríguez et al. 2011b). In the future genetic modification may form the basis for a strategy for pest and disease control. Also there is growing interest in breeding new cultivars with specific phytochemicals associated with improving their nutritional content to improve human health. It is possible to improve the antioxidant action of tomatoes by increasing the production of flavonoids by genetic modification, including the synthesis of specific flavonoids that are not produced naturally by tomatoes (Schijlen 2008). Tomato mutants have been isolated including non-ripening (*nor*) and ripening-inhibitor (*rin*), that do not produce climacteric ethylene. Genetic modification (GM) has produced tomatoes that do not soften normally by downregulation of the genes that control ethylene biosynthesis. These tomatoes can be harvested green and ripened in an atmosphere containing ethylene. One genetically engineered cultivar of tomato was called Flavr Savr and was marketed in the USA and to a limited extent as puree in Europe in the 1990s (Fig. 1.1). Due to subsequent restrictions in marketing products of biotechnology in the EU and some other countries, as well as some consumer resistance, the market was restricted and they were withdrawn from the EU. FlavrSavr had a "deactivated" gene, which is an antisense approach that meant that the plant was no longer able to produce polygalacturonase; an enzyme involved in fruit softening. The idea was that tomatoes could be left to ripen on the plant and still have a long shelf life thus allowing

Fig. 1.1 A tin of GM tomato
puree on sale in Britain in the
1990s

them to develop their full flavour since the best flavoured fruit were reported to be those that ripened fully on the plant (Chiesa et al. 1998) This approach came about because tomatoes are commonly harvested before they are fully ripe. The thinking behind this development was that when they are allowed to ripen fully on the plant they have become softer as a result of the natural ripening process. Since these fully ripe fruit are softer that those that are still green, or beginning to turn from green to red, they are more susceptible to damage during transport or they may become over ripe and unacceptable in the marketing chain.

Perhaps the best known application of GM technology to crops is the modification of soybeans to be resistant to the herbicide Glyphosate. These "Roundup Resistant" cultivars have completely changed the production of soybeans in countries like Argentina and the USA. GM can also be used to improve the quality of fruit and vegetables. For example, bananas have been modified to improve their nutritional benefits by genetically modifying them to improve their β-carotene content. GM has also been used to control pests for example nematodes in bananas.

Measurement and Control Technology

Temperature

The traditional way of measuring and controlling temperature within a store was the mercury or coloured alcohol glass thermometers. These were usually suspended from the ceiling and viewed through a window port in the door and the temperature control equipment adjusted manually. Because of the inconvenience and potential inaccuracy and unreliability of this method they have not been used commercially for many decades. Resistance thermometers, thermocouple or thermistors have replaced them. Resistance thermometers consist of a length of fine wire, typically made from platinum, nickel or copper, coiled around a ceramic or glass core and work by correlating the resistance of the element with temperature. Thermocouples consist of two dissimilar conductors that contact each other at one or more places to produce a voltage when the temperature of one of the places differs from the reference temperature at other parts of the circuit. Nickel–chromium positive conductors and nickel–aluminium negative conductors are both commonly used. Thermocouples are inexpensive and self powered but are not good for measurement in controlled atmosphere or hypobaric stores because their main limitation is accuracy since it is difficult to achieve less than 1 °C accuracy. Thermistors are temperature sensitive resistors that are inexpensive and are most commonly used in fruit and vegetable stores. All resistors vary with temperature, but thermistors are constructed of semiconductor material with a resistivity that is especially sensitive to temperature. This makes them very precise with an accuracy of ± 0.1 °C or ± 0.2 °C depending on the particular thermistor model. They retain this precision even through long cables that conduct the signal back to the recorders.

Humidity

Store humidity is referred to as percentage r.h. (humidity relative to that which is saturated) but it is also referred to as VPD (vapour pressure deficit) which relates the gaseous water in the atmosphere to its maximum capacity at a given temperature. VPD in air is the difference between the saturation vapour pressure and the actual vapour pressure at a given temperature. It can be expressed in millibars (mb) or millimetres of mercury (mm Hg). The vapour pressure is determined from the dry bulb and wet bulb readings by substitution in the following equation (Regnault, August and Apjohn quoted by Anonymous 1964):

$$e = e'_w - Ap(T - T')$$

where

e is the vapour pressure
e'_w is the saturation vapour pressure at the temperature T'
p is the atmospheric pressure
T is the temperature of the dry bulb
T' is the temperature of the wet bulb
A is a "constant" which depends on the rate of ventilation of the psychrometer, the latent heat of evaporation of water and the temperature scale in which the thermometers are graduated.

Low humidity in the storage atmosphere can result in initiation of ripening and thus a reduced storage life as well as increased water loss. Thompson et al. (1972) and Thompson et al. (1974a) described this effect on plantains. Maintaining high humidity around the fruit can help to keep fruit in the pre-climacteric stage so that where fruit were stored in moist coir dust or individual fingers were sealed in polyethylene film they can remain green and pre-climacteric for over 20 days in Jamaican ambient conditions. Such fruit ripened quickly and normally when removed.

Bishop (1990) commented "the author is unaware of any humidity sensor working on a long-term satisfactory basis in fruit stores" and "Most typical sensors do not specify performance in the high humidity band, some do but quote a ± 10 % or more error; while some claim better performance." For accurate non-recording measurements a dew point metre or a whirling hygrometer can be used, which are potentially very accurate so long as they are maintained and used correctly. The dew point is the temperature at which the water vapour in air at constant pressure condenses into liquid water at the same rate at which it evaporates. Equipment has been developed that uses dew point to measure humidity and can be used as a standard with which other methods may be calibrated. The practice is that water vapour condenses onto a temperature controlled mirror surface and the dew point is detected. Originally dew point metres were simply used for single spot readings. However, with the development of optical electronics several companies have metres that can be permanently installed or attached to battery operated display device with data recording capabilities. Whirling hygrometers consist of two mercury or alcohol in glass thermometer, one covered with a cotton sleeve that is kept wet and air is passed across the two thermometers and the depression of the temperature of one thermometer due to water evaporating from the sleeve compared to the one with no sleeve is proportional to the amount of moisture in the air and therefore to the dew point and humidity.

References

Aghdam, M.S. 2013. Role of alternative oxidase in postharvest stress of fruit and vegetables: chilling injury. *African Journal of Biotechnology* 12: 7009–7016.

Alighourchi, H., Barzegar, M., Abbasi, S., Zeitschrift fur Lebensmittel, U. and A. Forschung. 2008. Anthocyanins characterization of 15 Iranian pomegranate (*Punica granatum* L.) varieties and their variation after cold storage and pasteurization. *European Food Research and Technology* 227, 881–887.

Bastiaanse, H., de Lapeyre de Bellaire, L., Lassois, L., Misson, C., and M.H. Jijakli. 2010. Integrated control of crown rot of banana with *Candida oleophila* strain O, calcium chloride and modified atmosphere packaging. *Biological Control* 53, 100–107.

Berard, J.E. 1821. Memoire sur la maturation des fruits. *Annales de Chimie et de Physique* 16: 152–183.

Berard, L.S. 1985. Effects of CA on several storage disorders of winter cabbage. Controlled atmospheres for storage and transport of perishable agricultural commodities. *Fourth National Controlled Atmosphere Research Conference, July 1985. North Carolina State University, Raleigh, USA*. 150–159.

Berard, L.S., B. Vigier, and M.A. Dubuc Lebreux. 1986. Effects of cultivar and controlled atmosphere storage on the incidence of black midrib and necrotic spot in winter cabbage. *Phytoprotection* 67: 63–73.

Bishop, D. 1990. Controlled atmosphere storage. In *Cold and Chilled Storage Technology*, ed. C.J.V. Dellino, 66–98. London and Glasgow UK: Blackie and Sons Ltd.

Blackbourn, H.D., M.J. Jeger, P. John, and A.K. Thompson. 1990. Inhibition of degreening in the peel of bananas ripened at tropical temperatures, III changes in plastid ultrastructure and chlorophyll protein complexes accompanying ripening in bananas and plantains. *Annals of Applied Biology* 117: 147–161.

Blackman, F.F., and P. Parija. 1928. Analytic studies in plant respiration. I. The respiration of a population of senescent ripening apples. *Proceedings of the Royal Society, London Series B* 103: 412–445.

Blackman, F.F. 1928. Analytic studies in plant respiration. III. Formulation of a catalytic system for the respiration of apples and its relation to oxygen. *Proceedings of the Royal Society, London Series B* 103: 491–523.

Blauer, J.M., G.N. Mohan Kumar, L.O. Knowles, A. Dhingra, and N.R. Knowles. 2013. Changes in ascorbate and associated gene expression during development and storage of potato tubers (*Solanum tuberosum* L.). *Postharvest Biology and Technology* 78: 76–91.

Zhang, Bo, Wan-peng Xi, Wen-wen Wei, Ji-yuan Shen, Ian Ferguson, and Kun-song Chen. 2014. Changes in aroma-related volatiles and gene expression during low temperature storage and subsequent shelf-life of peach fruit. *Postharvest Biology and Technology* 60: 7–16.

Bufler, G. 1986. Ethylene-promoted conversion of 1-aminocyclopropane-1-carboxylic acid to ethylene in peel of apple at various stages of fruit development. *Plant Physiology* 80: 539–543.

Burg, S.P. 2004. *Postharvest physiology and hypobaric storage of fresh produce*. Wallingford, Oxford, UK: CAB International.

Chalutz, E., and M.A. Stahman. 1969. Induction of pisatin by ethylene. *Phytopathology* 59: 1972–1973.

Chalutz, E., E. Lomaniec, and J. Waks. 1989. Physiological and pathological observations on the post-harvest behaviour of kumquat fruit. *Tropical Science* 29: 199–206.

Chiesa, A., S. Moccia, D. Frezza, and S. Filippini de Delfino. 1998. Influence of potassic fertilization on the postharvest quality of tomato fruits. *Agricultura Tropica et Subtropica* 31: 71–78.

Crisosto, C.H. and Kader, A.A. 2002. Apricots. In *The Commercial Storage of Fruits, Vegetables, and Florist and Nursery Stocks*, ed. Gross K.C., Wan C.Y., and Saltveit M. USDA Agricultural Handbook Number 66. http://www.ba.ars.usda.gov/hb66/contents.html accesssed November 2012.

Curd, L. 1988. *The design and testing using strawberry fruits of an ethylene dilution system*. MSc thesis, Silsoe College, Cranfield Institute of Technology, UK.

D'Angelo, S., A. Cimmino, M. Raimo, A. Salvatore, V. Zappia, and P. Galletti. 2007. Effect of reddeningripening on the antioxidant activity of polyphenol extracts from cv. 'Annurca' apple fruits. *Journal of Agricultural and Food Chemistry* 55: 997–9985.

Dennis, C. 1983. Soft fruits. In *Postharvest pathology of fruits and vegetables*, ed. C. Dennis. UK: Academic Press London.

Desai, B.B., and P.B. Deshpande. 1975. Chemical transformations in three varieties of banana *Musa paradisica* Linn fruits stored at 20 °C. *Mysore Journal of Agricultural Science* 9: 634–643.

Du Pont, M.S., Z. Mondin, G. Williamson, and K.R. Price. 2000. Effects of variety, processing and storage on the flavenoid glycoside content and composition of lettuce and endive. *Journal of Agricultural and Food Chemistry* 48: 3957–3964.

Fahey, J.W., Y. Zhang, and P. Talalay. 1997. Broccoli sprouts: an exceptionally rich source of inducers of enzymes that protect against chemical carcinogens. *Proceedings of the National Academy of Sciences of the United States of America* 94: 10366–10372.

Ferrante, A., and T. Maggiore. 2007. Chlorophyll a fluorescence measurements to evaluate storage time and temperature of Valeriana leafy vegetables. *Postharvest Biology and Technology* 45: 73–80.

Ferris, R.S.B., H. Wainwright, and A.K. Thompson. 1993. Effect of maturity, damage and humidity on ripening of plantain and cooking banana. *ACIAR Proceedings* 50: 434–437.

Fidler, J.C. 1963. Refrigerated storage of fruits and vegetables in the U.K., the British Commonwealth, the United States of America and South Africa. *Ditton Laboratory Memoir* 93.

Fidler, J.C., B.G. Wilkinson, Edney, K.L. and R.O. Sharples. 1973. The biology of apple and pear storage. *Commonwealth Agricultural Bureaux Research Review,* 3, 235 pp.

Gane, R. 1934. Production of ethylene by some ripening fruits. *Nature* 134: 1008.

Goyette, B. 2010. Hyperbaric treatment to enhance quality attributes of fresh horticultural produce. PhD thesis, McGill University, Montreal, Canada.

Guadarrama, A., V. Ortiz, and J.P. Ogier. 2000. Postharvest comparative study of two cultivars of sapodilla (*Manilkara zapote* L.) fruits. *Acta Horticulturae* 536: 363–367.

Harb, J., Gapper, N.E., Giovannoni, J.J. and C.B. Watkins. 2012. Molecular analysis of softening and ethylene synthesis and signalling pathways in a non-softening apple cultivar, 'Honeycrisp' and a rapidly softening cultivar, 'McIntosh'. *Postharvest Biology and Technology* 64, 94–103.

Hasperué, J.H., M.E. Gómez-Lobato, A.R. Chaves, P.M. Civello, and G.A. Martínez. 2013. Time of day at harvest affects the expression of chlorophyll degrading genes during postharvest storage of broccoli. *Postharvest Biology and Technology* 82: 22–27.

Katz, E., P. Martinez-Lagunes, J. Riov, D. Weiss, and E.E. Goldschmidt. 2004. Molecular and physiological evidence suggests the existence of a system II-like pathway of ethylene production in non-climacteric Citrus fruit. *Planta* 219: 243–252.

Kidd, F., and C. West. 1925. The course of respiratory activity throughout the life of an apple. *Report of the Food Investigation Board London for* 1924: 27–34.

Kidd, F. and C. West. 1927. A relation between the concentration of O_2 and CO_2 in the atmosphere, rate of respiration, and the length of storage of apples *Report of the Food Investigation Board London for 1925, 1926*, 41–42.

Knee, M. 1973. Effects of controlled atmosphere storage on respiratory metabolism of apple fruit tissue. *Journal of the Science of Food and Agriculture* 24: 289–298.

Knee, M. 1976. Influence of ethylene on the ripening of stored apples. *Journal of the Science of Food and Agriculture* 27: 383–392.

Lafuente, M.T., M. Cantwell, S.F. Yang, and U. Rubatzky. 1989. Isocoumarin content of carrots as influenced by ethylene concentration, storage temperature and stress conditions. *Acta Horticulturae* 258: 523–534.

Lana, M.M., O. Van Kooten, M. Dekker, P. Suurs, and F.A. Linssen. 2005. Effects of cutting and maturity of lycopene concentration of fresh cut tomatoes during storage at different temperatures. *Acta Horticulturae* 682: 1871–1878.

Laval Martin, D., Quennemet, J. and R. Moneger. 1975. Remarques sur l'evolution lipochromique et ultrastructurales des plastes durant la maturation du fruit de tomate "cerise". Facteurs et regulation da la Maturation des Fruits. *Coll. Intern. du C.N.R.S.* 1974 374.

Lee, S.K. and S.D. Yun. 2002. Ginseng. In *The Commercial Storage of Fruits, Vegetables, and Florist and Nursery Stocks*, ed. Gross K.C., Wan C.Y. and Saltveit M. USDA Agricultural Handbook Number 66. http://www.ba.ars.usda.gov/hb66/contents.html Accessed November 2012.

Leja, M., A. Mareczek, A. Starzynska, and S. Rozek. 2001. Antioxidant ability of broccoli flower buds during short term storage. *Food Chemistry* 72: 219–222.

Manning, K. 1993. Regulation by auxin of ripening genes from strawberries, a non climacteric fruit. Postharvest Biology and Handling of Fruit, Vegetables and Flowers. *Meeting of the Association of Applied Biologists, London* 8 December 1993.

Massignan, L., R. Lovino, and D. Traversi. 1999. Trattamenti post raccolta e frigoconservazione di uva da tavola "biologica". *Informatore Agrario Supplemento* 55: 46–48.

McMurchie, E.J., W.B. McGlasson, and I.L. Eaks. 1972. Treatment of fruit with propylene gives information about the biogenesis of ethylene. *Nature* 237: 235–236.

Medlicott, A.P., and A.K. Thompson. 1985. Analysis of sugars and organic acids in ripening mango fruits *Mangifera indica* var. Keitt by high performance liquid chromatography. *Journal of the Science of Food Agriculture* 36: 561–566.

Napolitano, A., A. Cascone, G. Graziani, R. Ferracane, L. Scalfi, C. Di Vaio, A. Ritieni, and V. Fogliano. 2004. Influence of variety and storage on the polyphenol composition of apple flesh. *Journal of Agricultural and Food Chemistry* 52: 6526–6531.

Nuske, D., and H. Muller. 1984. Erste Ergebnisse bei der industriemassigen Lagerung von Kopfkohl unter CA Lagerungsbedingungen. *Nachrichtenblatt fur den Pflanzenschutz in der DDR* 38: 185–187.

Pavez, L., C. Hödar, F. Olivares, M. González, and V. Cambiazo. 2013. Effects of postharvest treatments on gene expression in *Prunus persica* fruit: Normal and altered ripening. *Postharvest Biology and Technology* 75: 125–134.

Pendergrass, A., and F.M.R. Isenberg. 1974. The effect of relative humidity on the quality of stored cabbage. *HortScience* 9: 226–227.

Perkins-Veazie, P., J.K. Collins, and L. Howard. 2008. Blueberry fruit response to postharvest application of ultraviolet radiation. *Postharvest Biology and Technology* 47: 280–285.

Rapisarda, P., M.L. Bianco, P. Pannuzzo, and N. Timpanaro. 2008. Effect of cold storage on vitamin C, phenolics and antioxidant activity of five orange genotypes [*Citrus sinensis* (L.) Osbeck]. *Postharvest Biology and Technology* 49: 348–354.

Rees, D. 2012. Introduction. In *Crop Post-Harvest: Science and Technology*, vol. 3, ed. Debbie Rees, Graham Farrell, and John Orchard, 1–4. Oxford, UK: Blackwell Science Ltd.

Ribeiro, R.A., F.L. Finger, M. Puiatti, and V.W.D. Casali. 2005. Chilling injury sensitivity in arracacha (*Arracacia xanthorrhiza*) roots. *Tropical Science* 45: 55–57.

Rodríguez, A., V. San Andrés, M. Cervera, A. Redondo, B. Alquézar, T. Shimada, J. Gadea, M.J. Rodrigo, L. Zacarias, L. Palou, M.M. Lopez, P. Castañeda, and L. Peña. 2011a. Terpene down-regulation in orange reveals the role of fruit aromas in mediating interactions with insect herbivores and pathogens. *Plant Physiology* 156: 793–802.

Rodríguez, A., V. San Andrés, M. Cervera, A. Redondo, B. Alquézar, T. Shimada, J. Gadea, M.J. Rodrigo, L. Zacarias, L. Palou, M.M. Lopez, P. Castañeda, and L. Peña. 2011b. The monoterpene limonene in orange peels attracts pests and microorganisms. *Plant Signalling and Behaviour* 11: 1–4.

Romero, I., M.T. Sanchez-Ballesta, R. Maldonado, M.I. Escribano, and C. Merodio. 2008. Anthocyanin, antioxidant activity and stress-induced gene expression in high CO_2-treated table grapes stored at low temperature. *Journal of Plant Physiology* 165: 522–530.

Salunkhe, D.K., and M.T. Wu. 1973. Effects of subatmospheric pressure storage on ripening and associated chemical changes of certain deciduous fruits. *Journal of the American Society for Horticultural Science* 98: 113–116.

Schijlen, E. 2008. http://www.pri.wur.nl/UK/newsagenda/news/Genetic_modification_a_tool_for_making_vegetables_and_fruit_even_healthier.htm accessed July 2008.

Sdiri, S., P. Navarro, A. Monterde, J. Benabda, and A. Salvador. 2012. Effect of postharvest degreening followed by a cold-quarantine treatment on vitamin C, phenolic compounds and antioxidant activity of early-season citrus fruit. *Postharvest Biology and Technology* 65: 13–21.

Seymour, G.B. 1985. *The effects of gases and temperature on banana ripening*. PhD thesis, University of Reading, UK.

Yan, Shijie, Ling Li, Lihua He, Liya Liang, and Xiaodan Li. 2013. Maturity and cooling rate affects browning, polyphenol oxidase activity and gene expression of 'Yali' pears during storage. *Postharvest Biology and Technology* 85: 39–44.

Shimada, T., T. Endo, H. Fujii, M. Hara, T. Ueda, M. Kita, and M. Omura. 2004. Molecular cloning and functional characterization of four monoterpene synthase genes from *Citrus unshiu* Marc. *Plant Science* 166: 49–58.

Shin, Y.R., A.L. Jung, H. Rui, J.F. Nock, and C.B. Watkins. 2008. Harvest maturity, storage temperature and relative humidity affect fruit quality, antioxidant contents and activity, and inhibition of cell proliferation of strawberry fruit. *Postharvest Biology and Technology* 49: 201–209.

Solomos, T., and J.B. Biale. 1975. Respiration in fruit ripening. *Colloque Internationaux du Center National de la Recherche Scientifique* 238: 221–228.

Somboonkaew, N., and L.A. Terry. 2011. Influence of temperature and packaging on physiological and chemical profiles of imported litchi fruit. *Food Research International* 44: 1962–1969.

Stevens, R., D. Page, B. Gouble, C. Garchery, D. Zamir, and M. Causse. 2008. Tomato fruit ascorbic acid content is linked with monodehydroascorbate reductase activity and tolerance to chilling stress. *Plant, Cell and Environment* 31: 1086–1096.

Suzuki, Y., and H. Okabayashi. 2001. Effect of temperature, humidity and packaging atmosphere on post-harvest disease of okra (*Hibiscus esculentus*) pods caused by *Alternaria alternata*. *Bulletin of the Kochi Agricultural Research Center* 10: 21–26.

Thompson, A.K. 2010. *Controlled atmosphere storage of fruits and vegetables*, 2nd ed. Oxford, UK: CAB International.

Thompson, A.K. 2015. *Fruit and vegetables - harvesting, handling and storage*, 3rd ed. Oxford, UK: Wiley Blackwell.

Thompson, A.K., and G.B. Seymour. 1982. Comparative effects of acetylene and ethylene gas on initiation of banana ripening. *Annals of Applied Biology* 101: 407–410.

Thompson, A.K., B.O. Been, and C. Perkins. 1972. Handling, storage and marketing of plantains. *Proceedings of the Tropical Region of the American Society of Horticultural Science* 16: 205–212.

Thompson, A.K., B.O. Been, and C. Perkins. 1974a. Effects of humidity on ripening of plantain bananas. *Experientia* 30: 35–36.

Thompson, A.K., B.O. Been, and C. Perkins. 1974b. Prolongation of the storage life of breadfruits. *Proceedings of the Caribbean Food Crops Society* 12: 120–126.

Toole, C.W. 1990. Bulk stores and associated services. In *Cold and Chilled Storage Technology*, ed. C.J.V. Dellino, 1–36. London, UK: Blackie.

Wan, C.K., and P.F. Lam. 1984. Biochemical changes, use of polyethylene bags, and chilling injury of carambola *Averrhoa carambola* L. stored at various temperatures. *Pertanika* 7: 39–46.

Wang, C.Y. 1990. Physiological and biochemical effects of controlled atmosphere on fruit and vegetables. In *Food Preservation by Modified Atmospheres*, ed. M. Calderon and R. Barkai-Golan, 197–223. Boca Raton, Ann Arbor, Boston, USA: CRC Press.

Wilkinson, B.G. 1972. Fruit storage. *East Malling Research Station Annual Report for* 1971: 69–88.

Wills, R.B.H., V.W. Ku, D. Shohet, and G.H. Kim. 1999. Importance of low ethylene levels to delay senescence of non-climacteric fruit and vegetables. *Australian Journal of Experimental Agriculture* 39: 221–224.

Wilson Wijeratnam, R.S., I.G.N. Hewajulige, and N. Abeyratne. 2005. Postharvest hot water treatment for the control of *Thielaviopsis* black rot of pineapple. *Postharvest Biology and Technology* 36: 323–327.

Winkler, S., J. Faraghera, P. Franza, M. Imsica, and R. Jonesa. 2006. Glucoraphanin and flavonoid levels remain stable during simulated transport and marketing of broccoli (*Brassica oleracea* var. *italica*) heads. *Postharvest Biology and Technology* 43: 89–94.

Wold, A.B., M. Hansen, K. Haffner, and R. Blomhoff. 2007. Antioxidant activity in tomato (*Lycopersicon esculentum* Mill.) and broccoli (*Brassica oleracea* var. *italica* L.) cultivars—effects of maturity and storage conditions. *Acta Horticulturae* 744: 381–386.

Xie, Xingbin, Jiankun Song, Yan Wang, and David Sugar. 2014. Ethylene synthesis, ripening capacity, and superficial scald inhibition in 1-MCP treated 'd'Anjou' pears are affected by storage temperature. *Postharvest Biology and Technology* 97: 1–10.

Yusuf, C., and B.A.O. Okusanya. 2008. Fungi associated with the storage rot of yam (*Dioscorea rotundata* Poir.) in Yola, Adamawa State. *Journal of Sustainable Development in Agriculture & Environment* 3: 99–103.

Zheng, Y., R.W. Fung, S.Y. Wang, and C.Y. Wang. 2008. Transcript levels of antioxidative genes and oxygen radical scavenging enzyme activities in chilled zucchini squash in response to superatmospheric oxygen. *Postharvest Biology and Technology* 47: 151–158.

Zhou, Y., J.M. Dahler, S.J.R. Underhill, and R.B.H. Wills. 2003. Enzymes associated with blackheart development in pineapple fruit. *Food Chemistry* 80: 565–572.

Chapter 2
Controlled Atmosphere Storage

Introduction

Clearly, the optimum temperature has an enormous effect on the postharvest like of fresh fruit and vegetables, but controlling the gaseous atmosphere in a store has been shown to improve the maintenance of their postharvest quality over and above the extensions gained by simply controlling the temperature and humidity. This chapter briefly reviews the effects of controlling the gases within the store and the technology that is used and puts the technology and effects in context with developments over time.

History

The effects of gases on harvested crops have been known for centuries. For example, Wang (1990) quotes a Tang dynasty eighth century poem that described how litchis were shown to keep better during long distance transport when they were sealed in the hollow centres of bamboo stems with some fresh leaves. The earliest documented scientific study of controlled atmosphere storage was by Berard (1821) who showed that fruit stored in atmospheres containing no O_2 did not ripen, but if they were held for only a short period and then placed in air they continued to ripen. In the 1850s and 1860s, a commercial cold storage company in the USA experimented with modifying the CO_2 and O_2 in an apple store by making it air tight. It was claimed that the apples were kept in good condition in the store for 11 months, but some fruit were injured; possibly by CO_2 toxicity (Dalrymple 1967). Some success was reported by Washington State University in the USA around 1903, and subsequently by others, on controlled atmosphere storage of apples, raspberries, blackberries, strawberries and loganberries. Sharples

© The Author(s) 2016
A.K. Thompson, *Fruit and Vegetable Storage*, SpringerBriefs in Food,
Health, and Nutrition, DOI 10.1007/978-3-319-23591-2_2

(1989) in his review in Classical Papers in Horticultural Science stated that "[Franklin] Kidd and [Cyril] West can be described as the founders of modern CA storage." Sharples described the background to their work and how it came about. Dalrymple (1967) in reviewing early work on the effects of gases on postharvest of fruit and vegetables stated "The real start of CA storage had to await the later work of two British scientists [Kidd and West], who started from quite a different vantage point". In 1918, the work being carried out at the Food Investigation Organisation in Cambridge was described as "a study of the normal physiology, at low temperatures, of those parts of plants which are used as food. The influence of the surrounding atmosphere, of its content of O_2, CO_2 and water vapour was the obvious point to begin at, and such work has been taken up by Dr. F. Kidd. The composition of the air in fruit stores has been suspected of being important and this calls for thorough elucidation. Interesting results in stopping sprouting of potatoes have been obtained, and a number of data with various fruits proving the importance of the composition of the air." (Anonymous 1919). Controlled atmosphere storage at low temperature of plums, apples and pears was described as "has been continuing" by Anonymous (1920) with large-scale gas storage tests on apples and pears. In 1920 a semi-commercial controlled atmosphere storage trial was set up at a farm at Histon in Cambridgeshire to test their laboratory findings in small scale commercial practice. In 1929 a commercial controlled atmosphere store for apples was built by a grower near Canterbury in Kent. Controlled atmosphere storage has continued to be used on an increasing scale, with an increasing variety of fruit and vegetables and with an increasing number of countries since that time (Thompson 2010).

Changes During Storage

The postharvest changes in fresh fruit and vegetables are affected by their postharvest environment as well as microorganism infection, the stage of their development or maturity at harvest and the conditions in which they have been grown. The changes also depend on the part of the plant or tree on which it has grown. Some vegetables are natural storage organs, for example potato tubers and onions bulbs, and their postharvest changes are different from say leaf vegetables, such as lettuce and cabbages. The postharvest requirements for fruit also can vary considerably. Fruits are often classified into climacteric and non-climacteric and, in some cases intermediate where their ripening metabolism is not clear. Climacteric fruits are those whose ripening is accompanied by an increase in respiration rate, called the climacteric rise, which is generally associated with increased ethylene production. Initiation of the climacteric rise in respiration rate is by ethylene biosynthesis and associated with other chemical and physical changes. There are more postharvest changes in climacteric fruit than in non-climacteric fruit or in vegetable. In non-climacteric fruit or vegetable, the chemical content remains similar during their postharvest life except for perhaps sugars which are utilised for metabolic processes and therefore decrease. In climacteric fruit, there are considerable changes that we commonly refer to as ripening where the fruit develops typical flavour and aroma, changes colour (through loss of chlorophyll and synthesis of carotenoids and other pigments), changes of starch into sugars and changes in the cell wall constituents that probably contribute to softening. Cell walls are complex structures composed of cellulose and pectin, derived from hexoses including glucose, galactose, rhamnose and mannose, as well as pentoses including xylose and arabinose and some of their derivatives including glucuronic acid and galacturonic acid. The changes in aroma volatile chemicals are important since they affect the acceptability of fruit and vegetables. Controlled atmosphere storage has been shown to suppress aroma production in apples, for example the aroma levels decreased during long-term storage with those stored in 1 % O_2 having the lowest rate of aroma production compared to those stored in air (Villatoro et al. 2008). However, Fellman et al. (2000) reported that aroma levels rapidly returned to normal when they were removed from the controlled atmosphere store. There are also many changes in phytochemicals that can affect their nutritional and health promoting characteristics.

There is only limited direct evidence on the effects of hypobaric or hyperbaric storage on many chemical and nutritional changes in fruit and vegetables (these will be discussed in subsequent chapters) but the effects of controlled atmosphere storage on their quality has been more comprehensive studied. A few examples are given as follows: Van Der Sluis et al. (2001) found that controlled atmosphere storage of apples did not affect antioxidant activity differently from storage in air. Leja and Ben (2003) found that anthocyanin content of apples did not decrease during controlled atmosphere storage. There were some differences between cultivars on the effects of controlled atmosphere storage on the chemical content of the apples tested. Forney et al. (2003) compared storage of blueberries at 0 °C either in air or a range of controlled atmosphere conditions from 1 to 15 % O_2 combined with 0–15 % CO_2 and found that total phenolics decreased by 5–16 %, total anthocyanins by 8–18 % and antioxidant capacity by 4–14 % during 9 weeks storage depending on the atmosphere. Patil and Shellie (2004) found that when ultra-low O_2 levels in storage was used as a quarantine treatment for grapefruit that the levels of ascorbic acid, lycopene and β-carotene were higher than the controls. For avocado fruit Meyer et al. (2011) stated that here was no information available on the effects of controlled atmosphere storage on health-related compounds.

The flavour of fruits is partly determined by their sugar and acid content. The sugar level in fully ripe apples is mainly determined by the proportion of starch to sugar at harvest since sugar losses due to fruit respiration is no more than 10 % (Knee and Sharples 1979). However, they found that acidity could fall by as much as 50 % during storage and that there was a good correlation between fruit acidity and sensory evaluation. Controlled atmosphere storage of apples in either 2 % O_2 + 98 % nitrogen or 2 % O_2 + 5 % CO_2 + 93 % nitrogen resulted in few organic volatile compounds being produced during the storage period (Hatfield and Patterson 1974). Even when the fruit were removed from storage, they did not synthesise normal amounts of esters during ripening and esters are a major component of their aroma and flavour. In apples and pears, butyl etha-noate, 2-methyl butyl ethanoate and hexyl ethanoate are typical flavour and aroma compounds that are synthesised during ripening while terpenoid compounds such as linalool, epoxide and α-farnesene have been shown to be synthesised in some apple cultivars (Dimick and Hoskins 1983). Leja and Ben (2003) found a large increase in total phenolics in Jonagold and Sampion apples during storage for 120 days at 1 °C in air or 2 % O_2 + 2 % CO_2 followed by 7 days at 16 °C. They actually found a slight decrease in anthocyanins during storage in air but not in the controlled atmosphere. In contrast, MacLean et al. (2006) detected no change in total phenolics of Red Delicious apples during storage for 120 days at 0–1 °C followed by 8 days at room temperature, but there was an increase in chlorogenic acid and a decrease in anthocyanins. In 'Rocha', pears stored for 4 months in 2 % O_2 + 0–5 % CO_2 at 2 °C. Goodenough and Thomas (1981) showed that tomatoes ripened in 5 % CO_2 + 5 % O_2 had suppressed chlorophyll degradation and sup-pressed synthesis of the carotenoids lycopene and xanthophyll. In apples reducing the O_2 level predominantly inhibited chlorophyll degradation and TA was high-est in 15 % O_2 + 10 % CO_2 and 5 % CO_2 + 3 % O_2 (Ben-Arie et al. 1993). Galvis-Sanchez et al. (2006) found no differences between storage atmospheres on the phytochemical content they measured in pears, but arbutin and flavan-3-ols increased while flavanols and hydroxycinnamic acid derivatives did not change in all atmospheres they studied. Martínez-Sánchez et al. (2006) found that the total flavonoid content of wild rocket (*Diplotaxis tenuifolia*) was approximately 100 mg 100 g^{-1} fresh weight and remained constant during storage or even increased at the end of the shelf-life in 5 % O_2 + 10 % CO_2. In contrast, it was degraded in those samples kept in air and the total content of vitamin C was higher in con-trolled atmosphere stored samples than those kept in air. A decrease in the total antioxidant capacity was observed during storage and it was particularly marked in samples stored in air.

Jeffery et al. (1984) showed that lycopene synthesis in tomatoes was sup-pressed during storage in 6 % CO_2 + 6 % O_2. Rogiers and Knowles (2000) stored four cultivars of Saskatoon Serviceberry (*Amelanchier alnifolia*) at 0.5 °C for 56 days in 2, 10 and 21 % O_2 factorially combined with 0.035 or 5 % CO_2. They found that the 5 % CO_2 atmosphere combined with either 21 or 10 % O_2 was most effective at minimising losses in fruit soluble solids, anthocyanin, firmness and weight. In blueberries, controlled atmosphere storage had little or no effect on phenolic content (Schotsmans et al. 2007). Zheng et al. (2003) found that total phenolics were increased in blueberries during storage at 5 °C in 60–100 % O_2 for 35 days to a greater extent than those stored in air or 40 % O_2. In grapes, anthocy-anin levels were lower after storage at 0 °C for those that had been pre-treated for 3 days in 20 % CO_2 + 20 % O_2 compared to those that had not been pre-treated (Romero et al. 2008). Storage of snow pea pods in either 2.5 % O_2 with 5 % CO_2 or 10 % CO_2 with 5 % O_2 concentrations resulted in the development slight off-flavours, but this effect was reversible since it was partially alleviated after ventila-tion (Pariasca et al. 2001).

Damage

When the O_2 level in storage is too low or the CO_2 level too high, the crop can be damaged. Fidler et al. (1973) reported that the appearance of CO_2 injury symptoms is a function of concentration, exposure time and temperature. They describe external CO_2 injury in apples where "initially the damaged area is markedly sunken, deep green in colour and with sharply defined edges. Later in storage the damaged tissue turns brown and finally almost black". Injury caused as a result of low O_2 levels is due to fermentation resulting in the accumulation of toxic products usually alcohols and aldehydes, which can result in necrotic tissue that tends to begin at the centre of the fruit. The lower O_2 limit for apples was found to vary between cultivar from a low of about 0.8 % for Northern Spy and Law Rome to a high of about 1.0 % for McIntosh in cold storage. For blueberries, the lower O_2 limit increased with temperature and CO_2 level. Raising the temperature from 0 to 25 °C caused the lower O_2 limit to increase from about 1.8 % to approximately 4 %. Raising CO_2 levels from 5 to 60 % increased the lower O_2 limit for blueberry fruits from approximately 4.5 to >16 % (Beaudry and Gran 1993). Wardlaw (1938) showed that high CO_2 can cause surface-scald browning, pitting and excessive decay in aubergines and these symptoms are similar to those caused by chilling injury. Mencarelli et al. (1989) described CO_2 injury of aubergines as external browning without tissue softening and showed that susceptibility to CO_2 varied between cultivars. Gadalla (1997) showed that onions stored in 10 % CO_2 developed internal browning.

Residual Effects

There is considerable evidence in the literature that storing fruits and vegetables in CA storage can affect their subsequent shelf or marketable life (Thompson 2010). For example, Bell peppers exposed to 1.5 % O_2 for 1 day exhibited suppressed respiration rate for at least 24 h after transfer to air (Rahman et al. 1993). Burdon et al. (2008) showed that avocados that had been stored in controlled atmospheres had a longer shelf-life than those that had been stored in air for a similar period. Khanbari and Thompson (1996) showed that potatoes in controlled atmosphere storage did not sprout either during storage or when they had been removed. Wills et al. (1982) showed that pre-climacteric bananas exposed to low O_2 took longer to ripen when subsequently exposed to air than fruits kept in air for the whole period.

Measurement and Control Technology

Temperature and humidity are controlled in controlled atmosphere stores in the same as those described in Chap. 1. This section therefore deals only with O_2, CO_2 and ethylene.

Carbon Dioxide and Oxygen

The original way, and one that is still in common use, to control the CO_2 and O_2 levels in a controlled atmosphere store was by constant analysis. In many systems, the level of O_2 was allowed to reduce by sealing the room and allowing O_2 level to reduce by the respiration of the fruit. When the required level was reached, it was maintained at that level by frequently introducing fresh air from outside of the store. Usually tolerance limits were set at, say, plus or minus 0.1 % so that, if say 1 % O_2 was required, when the O_2 level went down to 0.9 % air was vented until it reached 1.1 %. CO_2 level in a store will increase, again through fruit respiration, and when it reaches the required level it is removed by passing the store air through or past a chemical that will remove the CO_2 and return the air back into the store. This method of CO_2 removal is called 'active scrubbing'. Alternatively, the CO_2 removing chemical may be placed inside the store where it can keep the level generally at low levels (usually about 1 %). This method is called 'passive scrubbing'. These methods of controlling O_2 and CO_2 in controlled atmosphere stores are referred to as 'product generated', since the gas levels are produced by the crops' respiration. The time taken for the levels of these two gases to reach the optimum (especially for the O_2 to fall from the 21 % in normal air) can reduce the maximum storage life of the crop. It is common therefore to fill the store with the crop, seal the store and inject nitrogen gas until the O_2 has reached the required level and then maintain it in the way described above. Scrubbers to control CO_2 are generally classified according to the mode of absorption (i.e. chemical or physical), or to the mode of air passage through the absorbing agent. Material used in chemical removal systems includes calcium hydroxide, sodium hydroxide, zeolites (alumino-silicate minerals) and activated charcoal. Hydroxides react irreversibly with the CO_2 producing carbonates. These must be replaced by fresh hydroxides when the reaction is complete. Bishop (1990) calculated that 1 kg of calcium hydroxide will adsorb 0.59 kg of CO_2 before it needs to be replaced. Koelet (1992) calculated that for one tonne of apples, 7.5 kg of calcium hydroxide was needed every 6–10 weeks depending on which cultivar was stored. Gas removal using zeolites and activated charcoal is based on the fixing of CO_2 in a particular way, and then releasing it again on contact with the outside air. So for this method, two stage systems have been developed where store air is passed through one part of the equipment while the other part is being ventilated by fresh air. The system is then reversed and so on.

The atmosphere in many modern controlled atmosphere stores is constantly analysed for CO_2 and O_2 levels using an infra-red gas analyser to measure CO_2 and a paramagnetic analyser for O_2. The analysers are monitored and controlled by a computer. In early controlled atmosphere stores, an Orsat gas analyser was used. This ingenious analyser was patented around 1873 by H Orsat and was used by taking a sample of gas through a valve in the store wall, which was then pumped to consecutive absorption bottles where CO_2 and O_2 were absorbed separately. After absorption of the CO_2 and O_2, the volume of the remaining gas mixture was verified, thus allowing determination of volumes. The Orsat gas analyser consists of a calibrated water-jacketed gas burette connected by glass capillary tubing to two absorption pipettes, one containing potassium hydroxide solution to absorb CO_2 and the other potassium pyrogallate solution to absorb O_2. By means of a rubber tubing arrangement, the gas to be analysed was drawn into the burette and flushed through several times. Typically, 100 ml was withdrawn for ease of calculation. Using the stopcocks that isolate the absorption burettes, the level of gas in the levelling bottle and the burette was adjusted to the zero point of the burette. The gas was then passed into the potassium hydroxide burette, left to stand for about two minutes and then withdrawn, isolating the remaining gas via the stopcock arrangements. The process was repeated to ensure full absorption. After levelling the liquid in the bottle and burette, the remaining volume of gas in the burette indicates the percentage of CO_2 absorbed. The same technique was repeated for O_2, using the potassium pyrogallate. A 100 ml gas sample will give about 0.1 % resolution. The volume of a gas, of course, varies with temperature and pressure and therefore these variables are need to be corrected. Some considerable skill was involved in making accurate measurement and in the 1970s, one of the Experimental Officers at the Tropical Products Institute in London (the late Peter Crowther) was very skilled with an Orsat and could achieve a much higher resolution then the rest of us.

The benefits of O_2 levels as low as 1 %, or even less, have been shown in extending the storage of some fruits; for example Table 2.1 shows the progressive extension in the storage life of apples over the years mainly due to lower O_2 levels in store. Very accurate control of O_2 level at these very low concentrations is vital in order not to damage the fruit. Methods that have been developed are based on approaches to the physiology of the fruit. There are three main approaches: one based on respiratory quotient (RQ) one based on ethanol biosynthesis and one based on chlorophyll fluorescence. Wollin et al. (1985) discussed the possibility that RQ may be used to calculate the lowest oxygen level that can be tolerated in fruit storage to be incorporated in an automated system. International Controlled Atmosphere Limited developed a system called 'Safepod' to measure the CO_2 and O_2 and calculate RQ within a sample chamber. The Safepod sits in the controlled atmosphere storage room and thus has the same temperature, humidity, pressure

Table 2.1 Changes in the recommended storage conditions for cox's orange pippin apples all at 3.5 °C (Bishop 1994)

O_2 %	CO_2 %	Approximate storage time in weeks	Approximate date of implementation
21	0	13	–
16	5	16	1920
3	5	21	1935
2	<1	27	1965
1¼	<1	31	1980
1	<1	33	1986

and atmosphere as the store. Periodically the valves are closed and the CO_2 and O_2 and RQ are measured. The Van Amerongen/AgroFresh uses RQ by measuring O_2 and CO_2 in stores feeding the data into a computer, which initiates an alarm at a pre-determined RQ level.

Where these very low O_2 levels were used in commercial controlled atmosphere stores in the 1990s an alcohol detector was fitted which sounded an alarm if ethanol fumes were detected as a result of fermentation in the fruit. Fermentation in fruit occurs when the O_2 level is insufficient to support the oxidative chemical processes in fruit and vegetables. Where fermentation (anaerobic respiration) begins, it is called the anaerobic compensation point. This anaerobic compensation point varies with type and cultivar of fruit as well as their physiological maturity and storage conditions. When the detector alarm sounded the store operator could increase the O_2 level and, where this was done quickly, no damage was done to the fruit. This technology was subsequently developed and Schouten et al. (1997) described a system which he called "dynamic control of ultra-low oxygen storage" based on headspace analysis of ethanol levels that were maintained at less than 1 ppm. With an alarm in place O_2 levels as low as 0.3–0.7 % could be maintained in the store. Computer controls were subsequently developed for this system. Schouten et al. (1997) described storage of the apple cultivar Elstar with the ethanol level in the store maintained below 1 ppm in an atmosphere of 0.3–0.7 % O_2 + < 0.5 % CO_2 that retained fruit quality better than those stored in 1.2 % O_2 + 2.5 % CO_2.

Recently, other stresses associated with metabolic responses of fruit and vegetables to low O_2 levels have been developed, called dynamic controlled atmosphere (DCA) or dynamic controlled atmosphere-chlorophyll fluorescence (DCA-CF). A link between the minimum fluorescence (*Fo*) and a metabolic shift from predominantly aerobic to fermentative metabolism (the lower O_2 limit) is the foundation of DCA (Wright et al. 2012). One method has been developed and was patented in Canada in 2001 as HarvestWatch[TM] (Prange et al. 2002; DeLong et al. 2004). HarvestWatch[TM] uses a computer programme that can automatically adjust the O_2 level when stress, based on chlorophyll fluorescence measurement, is detected. DCA storage requires leak-proof capacity of 0.1 m^2 100 m^{-3} or less. Prange et al. (2014) used DCA-CF to calculate the lower oxygen limit for apples and showed that this reduced considerably during storage for three of the cultivars tested (Table 2.2). They also found that this DCA-CF system was sensitive to other stresses that can occur in fruit during storage including CO_2 toxicity, chilling injury, 1-methylcyclopropene treatment, toxic ammonia refrigeration gas and desiccation as well as lower oxygen limit.

Various commercial systems have been developed including Isolcell and Storex. The Isolcell system is a commercial application of the Harvest Watch[TM] chlorophyll fluorescence method incorporated into Isolcell's atmosphere control equipment and installed in some 2000 commercial CA stores since 2003. Fruit samples are placed in 'kennels' within the CA store where the chlorophyll fluorescence is closely monitored by the computerised control system which enables corrective action to be taken when low O_2 levels are detected in the fruit. The first large-scale DCA installation was completed in the UK in 2013 by Isolcell in conjunction with UKCA Ltd. The Storex (DCS) system is based on ethanol

Table 2.2 The effects of time in storage on the lower oxygen limit detected by Dynamic Controlled Atmosphere-Chlorophyll Fluorescence (DCA-CF) on four apple cultivars (Prange et al. 2014)

Apple cultivar	Lower oxygen limit	
	10–19 October (%)	1–4 December (%)
Delicious	0.85	0.47
Golden delicious	0.92	0.45
Honeycrisp	0.90	0.50
Empire	0.90	0.88

production in fruit and uses smaller sample chambers integrated into the main store enabling low level measurements of ethanol from the fruit samples when the anaerobic compensation point is reached. Storex have installed their DCS system in commercial stores in Holland (http://www.ukcaltd.com/ accessed April 2015).

Ethylene

The measurement of ethylene in the laboratory can be carried out using a gas chromatograph fitted with a flame ionisation detector. Detector tubes are used in packhouses and stores. These are filled with molybdate palladium reagent and the most sensitive will indicate 0.5–10 $\mu L\ L^{-1}$ ethylene concentration. Ethylene can be measured successfully with a portable gas chromatograph fitted with a photo ionisation detector capable of measuring ethylene to a concentration below 0.01 uL L^{-1}. EASI-1 uses a proprietary 'nanoporous gold sensor technology' for "accurate real-time measurement of ethylene gas concentrations". This is licensed from Fluid Analytics in USA, which offers a claimed sensitivity to ethylene in the air at levels as low as 10 ppb. Levels of ethylene in the atmosphere due to pollution were measured by Lawton (1991), which showed that levels were very low with a maximum of 0.038 $\mu L\ L^{-1}$ (Table 2.3). Ethylene levels measured in a packhouse were higher than in a store for kiwifruit due to the engines in the forklifts, while in stores levels were considerably higher, especially in CA stores (Table 2.4).

Table 2.3 Ethylene levels in ambient air in $\mu L\ L^{-1}$. Modified from Lawton 1991

Sample locations	Ethylene concentration
Australian terminals	0–0.015
New Zealand terminals	0–0.026
New Zealand fruit terminals	0.002–0.038
Belgium fruit terminals	0.003–0.015
Pacific ocean	0–0.009
Atlantic ocean	0–0.010

Table 2.4 Ethylene concentrations in fruit stores $\mu L\ L^{-1}$. Modified from Lawton 1991

Sample locations	Ethylene concentration	Source of ethylene
Kiwifruit packhouse	0–0.070	Forklifts
Kiwifruit stores	0.005–0.055	Fruit
Air apple stores	1–30	Fruit
Air pear stores	2–25	Fruit
CA pear store	11–118	Fruit
CA apple store	27–243	Fruit

As will be described later, one of the additional benefits of hypobaric storage is the removal of ethylene from the store and even from individual fruit or vegetable cells. Ethylene is naturally produced by plant cells and for climacteric fruit it is responsible for initiating the ripening process. Exposure to ethylene can also cause negative effects, for example chlorophyll breakdown resulting in degreening and leaf abscission of leafy vegetables. Another negative effect can occur in mushrooms where exposure to ethylene can stimulate the stalk to elongate and the cap to expand and kiwifruit soften significantly during storage at 0 °C in response to ethylene concentrations as low as 10 nl L^{-1} (Retamales and Campos 1997). The major effect of ethylene removal in apple stores was shown to delay in the onset of softening and also slow softening once it has started (Dover and Stow 1993). In persimmons, exposure to 1 and 10 $\mu l\ L^{-1}$ ethylene at 20 °C also accelerated softening and limited their marketability therefore ethylene removal or exclusion during transport and storage was recommended by Crisosto et al. (1995).

Controlled atmosphere storage can reduce or eliminate detrimental effects of ethylene accumulation possibly by the increased levels of CO_2 competing for sites of ethylene action within the cells of the fruit. Stow et al. (2000) studied the effects of ethylene in controlled atmosphere apple stores and concluded that to obtain a benefit from ethylene removal, internal ethylene concentrations must be kept below about 4 $\mu l\ m^{-3}$ (0.1 ppm). The control of internal concentrations of ethylene in crops may be ultimately limited by the resistance of the crop to diffusion rather than its removal from the atmosphere surrounding the crop (Dover and Stow 1993). Tubamet AG reported that when they placed their Swingtherm ethylene absorber in cold stores the levels of ethylene were reduced. These reductions varied but were measured as 0.05 ± 0.1 ppm ethylene in citrus, pear and vegetable cells and <0.02 ppm ethylene for kiwifruit cells. Scald, a physiological storage disorder of apples and pears, has been associated with ethylene levels in the store atmosphere. Scald can be controlled by a pre-storage treatment with a suitable chemical antioxidant such as 1,2-dihydro-2,2,4-trimethylquinoline-6-yl ether or diphenylamine but there is consumer pressure to reduce postharvest chemical treatment and reducing ethylene in store may be an effective alternative to these chemical treatments. Coquinot and Richard (1991) stored apples in an atmosphere containing 1.2 % O_2 and 1 % CO_2 with or without removal of ethylene and found that in this atmosphere scald was controlled and ethylene removal was not necessary.

Baumann (1989) described a simple scrubber system which could be used in stores to remove both CO_2 and ethylene using activated charcoal. He gave a chart that showed the amount of activated charcoal required in relation to the CO_2 levels required and ethylene output of the fruit in the store. Molecular sieves and activated carbon can hold CO_2 and organic molecules such as ethylene. When fresh air is passed through these substances, the molecules are released. This means that they can be used in a two stage system where the store air is being passed through the substance to absorb the ethylene, while the other stage is being cleared by the passage of fresh air. After an appropriate period, the two stages are reversed. Hydrated aluminium silicate or aluminium calcium silicate are used. The regeneration of the molecular sieve beds can be achieved when they are warmed to 100 °C to drive off the CO_2 and ethylene. This system of regeneration is referred to as 'temperature swing' where the gases are absorbed at low temperature and released at high temperature. Two types of ethylene scrubber are marketed by UKCA Ltd where the store air is passed through a chamber either containing a hot metal catalyst (manufactured by Absoger, France) or alternatively relatively high concentrations of Ozone (Bioturbo[TM] manufactured by Miatech, USA). Both systems are also reported to kill airborne pathogens and it has been claimed that they can kill over 99.5 % of airborne bacteria and fungal spores thus contributing to the control of postharvest diseases on the fruit or vegetables. (http://www.ukcaltd.com/ accessed April 2015). However, ozone is hazardous to human health and is highly corrosive which can result in damage to storage facilities. The Bioturbo[TM] system overcomes these two problems by containing the whole process inside a discrete unit so that ozone is contained in the unit and is never released into the store atmosphere. This also means that it can be used at higher concentrations, which greatly increases effectiveness. Both these systems process the returning air so that it is fit to reintroduce to the storage chamber or packhouse.

Catalytic converters remove ethylene by chemical reaction. Air from the store is passed through a device where it is heated to over 200 °C in the presence of an appropriate catalyst, usually platinum (Wojciechowski 1989). Under these conditions, the ethylene in the air is oxidised to carbon dioxide and water. It requires an energy input of 30–80 watts per cubic metre of purified air, so it is a high energy consuming method. However, with suitable heat exchanges it is possible to make the method more energy efficient. One such device, called 'Swingtherm', reduced energy consumption to 7–14 W m^{-3}. Another ethylene converting device was marketed by Tubamet AG of Vaduz in Liechtenstein in 1993 and called "Swingcat". They took out a patent (Serial Number: 74095681) for a heated catalyst scrubber for the elimination of organic air pollutants. Portable ethylene scrubbers are available that can be placed in a store or packhouse (Fig. 2.1).

Chemicals can be used to remove or absorb ethylene. Proprietary products, including Ethysorb®, Purafil®, Consever-21 and Bi-On 4 are available which are basically made by impregnating an active alumina or zeolite clay carrier with a saturated solution of potassium permanganate and then drying it. Any molecule of ethylene in the atmosphere that comes into contact with the granule will be oxidised, therefore they are formed into small granules; the smaller the granules, the

Fig. 2.1 Tubamet AG
Swingcat portable ethylene
scrubber installed in a cold
store in UK

larger the surface area and therefore quicker their absorbing characteristics. The oxidising reaction is not reversible and the granules change colour from purple to brown which indicates that they need replacing. Strop (1992) studied the effects of storing broccoli in PE film bags with and without Ethysorb. She found that the ethylene content in the bags after 10 days at 0 °C was 0.423 $\mu l\ L^{-1}$ for those without Ethysorb and 0.198 $\mu l\ L^{-1}$ for those with Ethysorb. However, Scott et al. (1971) showed that the inclusion of potassium permanganate in sealed packages reduced the mean level of ethylene from 395 to 1.5 $\mu l\ L^{-1}$. Where potassium permanganate was included in the bags containing bananas the increase in storage life was 3–4 times compared to non-wrapped fruit and could be stored for 6 weeks at 20 or 28 °C and 16 weeks at 13 °C (Satyan et al. 1992). Kiwifruit are very susceptible to ethylene in the storage atmosphere. Ben-Arie and Sonego (1985) found that kiwifruit stored in sealed polyethylene film bags containing Ethysorb had less than 0.01 $\mu l\ L^{-1}$ ethylene resulting in a slower rate of softening and improved keeping compared to those with no ethylene absorbent. Terry et al. (2007) described a palladium-impregnated zeolite giving finely dispersed palladium particles that was far superior to potassium permanganate-based scavengers when used in low amounts.

Lawton (1991) evaluated four techniques for the removal of ethylene in ambient air and cargoes prior to and during refrigerated transport from the southern hemisphere to Europe. The methods were: ventilation with air, potassium permanganate, platinum catalyst heated to approximately 250 °C and ultraviolet radiation at 184 and 254 nm. He concluded that ventilation with air was the best method for the removal of ethylene gas. In the holds of a ship recently loaded with New Zealand kiwifruit, ethylene gas concentrations were found to be low, between 0.001 and 0.008 μl L^{-1} (Lawton 1991). In an extensive review Keller et al. (2013) concluded that photo-catalysis offered the greatest potential for removing ethylene. Photo-catalytic oxidation is a combination of a catalyst (usually titanium dioxide, but other catalysts have been used) and light that breaks down volatile organic compounds such as ethylene into carbon dioxide and water. Lin et al. (2013) reported that intermediates have been detected in some photo-catalytic oxidation processes that can poison the active sites resulting in deactivation of catalysts as well as being more toxic to human health. These should be removed or further oxidised to CO_2.

As indicated above, ethylene reduction or removal from fruit and vegetable stores is beneficial. However, with papaya Broughton et al. (1977) showed that scrubbing ethylene from a cold store had no effects on their storage life. In a study of ethylene on pears, Retamales et al. (1998) found little benefit in removing ethylene from pears during storage at −0.5 °C and Bower et al. (2003) concluded that although it is desirable to minimise ethylene in the storage atmosphere for pears, benefits are likely to be minor compared with the potential gains from good temperature management.

References

Anonymous 1919. Food Investigation Board. Department of Scientific and Industrial Research Report for the Year, 1918.

Anonymous 1920. Food Investigation Board. Department of Scientific and Industrial Research Report for the Year, 1920, 16–25.

Baumann, H. 1989. Adsorption of ethylene and CO_2 by activated carbon scrubbers. *Acta Horticulturae* 258: 125–129.

Beaudry, R.M., and C.D. Gran. 1993. Using a modified-atmosphere packaging approach to answer some post-harvest questions: Factors influencing the lower O_2 limit. *Acta Hortciulturae* 326: 203–212.

Ben-Arie, R., and L. Sonego. 1985. Modified-atmosphere storage of kiwifruit (*Actinidia chinensis* Planch) with ethylene removal. *Scientia Horticulturae* 27: 263–273.

Ben-Arie, R., A. Levine, L. Sonego, and Y. Zutkhi. 1993. Differential effects of CO_2 at low and high O_2 on the storage quality of two apple cultivars. *Acta Horticulturae* 326: 165–174.

Berard, J.E. 1821. Memoire sur la maturation des fruits. *Annales de Chimie et de Physique* 16: 152–183.

Bishop, D. 1990. Controlled atmosphere storage. In *Cold and Chilled Storage Technology*, ed. C.J.V. Dellino, 66–98. London, UK: Blackie.

Bishop, D.J. 1994. Application of new techniques to CA storage. *Commissions C2, D1, D2/3 of the International Institute of Refrigeration International Symposium* June 8–10 Istanbul Turkey, 323–329.

Bower, J.H., W.V. Biasi, and E.J. Mitcham. 2003. Effect of ethylene in the storage environment on quality of 'Bartlett pears'. *Postharvest Biology and Technology* 28: 371–379.

Broughton, W.J., A.W. Hashim, T.C. Shen, and I.K.P. Tan. 1977. Maturation of Malaysian fruits. I. Storage conditions and ripening of papaya *Carica papaya* L. cv. Sunrise Solo. *Malaysian Agricultural Research and Development Institute Research Bulletin* 5: 59–72.

Burdon, J., N. Lallu, G. Haynes, K. McDermott, and D. Billing. 2008. The effect of delays in establishment of a static or dynamic controlled atmosphere on the quality of 'Hass' avocado fruit. *Postharvest Biology and Technology* 49: 61–68.

Coquinot, J.P., and L. Richard. 1991. Methods of controlling scald in the apple Granny Smith without chemicals. Methodes de controle de l'echaudure de la pomme Granny Smith sans adjuvants chimiques. *Neuvieme colloque sur les recherches fruitieres, 'La maitrise de la qualite des fruits frais', Avignon*, 4–5–6 Decembre 1990, 373–380.

Crisosto, C.H., Mitcham, E.J., and A.A. Kader. 1995. Produce facts: Persimmons. *Perishables Handling* 84: 19–20. http://postharvest.ucdavis.edu/produce/producefacts/index.html.

Dalrymple, D.G. 1967. The development of controlled atmosphere storage of fruit. Division of Marketing and Utilization Sciences, Federal extension Service, U.S. Department of Agriculture.

DeLong, J.M., R.K. Prange, J.C. Leyte, and P.A. Harrison. 2004. A new technology that determines low-oxygen thresholds in controlled-atmosphere-stored apples. *HortTechnology* 14: 262–266.

Dimick, P.S., and J.C. Hoskins. 1983. Review of apple flavour. State of the Art. *CRC Critical Review, Food Science and Nutrition* 18: 387–409.

Dover, C.J., and J.R. Stow. 1993. The effects of ethylene removal rate and period in a low ethylene atmosphere on ethylene production and softening of Cox apples. *Sixth Annual Controlled Atmosphere Research Conference, Cornell University, Ithica, New York.*

Fellman, J.K., T.W. Miller, D.S. Mattinson, and J.P. Mattheis. 2000. Factors that influence biosynthesis of volatile flavour compounds in apple fruits. *HortScience* 35: 1026–1033.

Fidler, J.C., B.G. Wilkinson, K.L. Edney, and R.O. Sharples. 1973. The biology of apple and pear storage. *Commonwealth Agricultural Bureaux Research Review 3.*

Forney, C.F., M.A. Jordan, and K.U.K.G. Nicholas. 2003. Effect of CO_2 on physical, chemical, and quality changes in 'Burlington' blueberries. *Acta Horticulturae* 600: 587–593.

Gadalla, S.O. 1997. Inhibition of sprouting of onions during storage and marketing. *PhD thesis, Cranfield University, UK.*

Galvis-Sanchez, A.C., F.C. Fonseca, A. Gil-Izquierdo, M.I. Gil, and F.X. Malcata. 2006. Effect of different levels of CO_2 on the antioxidant content and the polyphenol oxidase activity of 'Rocha' pears during cold storage. *Journal of the Science of Food and Agriculture* 86: 509–517.

Goodenough, P.W., and T.H. Thomas. 1981. Biochemical changes in tomatoes stored in modified gas atmospheres. I. sugars and acids. *Annals of Applied Biology* 98: 507.

Hatfield, S.G.S., and B.D. Patterson. 1974. Abnormal volatile production by apples during ripening after controlled atmosphere storage. In: *Facteurs and Regulation de la Maturation des Fruits. Colleques Internationaux, CNRS, Paris*, 57–64.

Jeffery, D., C. Smith, P.W. Goodenough, T. Prosser, and D. Grierson. 1984. Ethylene independent and ethylene dependent biochemical changes in ripening tomatoes. *Plant Physiology* 74: 32.

Keller, N., M.N. Ducamp, D. Robert, and V. Keller. 2013. Ethylene removal and fresh product storage: A challenge at the frontiers of chemistry. Toward an approach by photocatalytic oxidation. *Chemical Review* 113: 5029–5070.

Khanbari, O.S., and A.K. Thompson. 1996. Effect of controlled atmosphere, temperature and cultivar on sprouting and processing quality of stored potatoes. *Potato Research* 39: 523–531.

Knee, M., and R.O. Sharples. 1979. Influence of CA storage on apples. In: Quality in stored and processed vegetables and fruit, *Proceedings of a Symposium at Long Ashton Research Station, University of Bristol 8–12 April 1979*, 341–352.

Koelet, P.C. 1992. *Industrial Refrigeration*. London, UK: MacMillan.

Lawton, A.R. 1991. Measurement of ethylene gas prior to and during transport. *19th International Congress of Refrigeration, IIR/IIF, Montreal*, 1–11.

Leja, M., and J. Ben. 2003. Antioxidant properties of two apple cultivars during long term storage. *Food Chemistry* 80: 303–307.

Lin, Lin, Yuchao Chai, Bin Zhao, Wei Wei, Dannong He, Belin He, and Qunwei Tang. 2013. Photocatalytic oxidation for degradation of Volatile organic compounds. *Open Journal of Inorganic Chemistry* 3: 14–25.

MacLean, D.D., D.P. Murr, J.R. DeEll, and C.R. Horvath. 2006. Postharvest variation in apple (*Malus* × *domestica* Borkh.) flavonoids following harvest, storage, and 1-MCP treatment. *Journal of Agriculture and Food Chemistry* 54: 870–878.

Martínez-Sánchez, A., A. Allende, R. Bennett, F. Ferreres, and M. Gil. 2006. Microbial, nutritional and sensory quality of rocket leaves as affected by different sanitizers. *Postharvest Biology and Technology* 42: 86–97.

Mencarelli, F., Fontana, F., and R. Massantini. 1989. Postharvest practices to reduce chilling injury CI on eggplants. Proceedings of the fifth international controlled atmosphere research conference, Wenatchee, Washington, USA, 14–16 June, 1989. Vol. 2, Pullman, Washington, USA, Washington State University, pp. 49–55.

Meyer, M.D., S. Landahl, M. Donetti, and L.A. Terry. 2011. Avocado. In *Health-promoting properties of fruit and vegetables*, ed. L.A. Terry, 27–50. Oxford, UK: CAB International.

Pariasca, J.A.T., T. Miyazaki, H. Hisaka, H. Nakagawa, and T. Sato. 2001. Effect of modified atmosphere packaging (MAP) and controlled atmosphere (CA) storage on the quality of snow pea pods (*Pisum sativum* L. var. *saccharatum*). *Postharvest Biology and Technology* 21: 213–223.

Patil, B.S., and K. Shellie. 2004. Carotenoids and vitamin C changes by semi-commercial ultralow oxygen storage in grapefruit. *Acta Horticulturae* 632: 321–328.

Prange, R.K., J.M. Delong, J.C. Leyte, and P.A. Harrison. 2002. Oxygen concentration affects chlorophyll fluorescence in chlorophyll-containing fruit. *Postharvest Biology and Technology* 24: 201–205.

Prange, R.K., Wright, A.H., DeLong, J.M., and A. Zanella. 2014. History, current situation and future prospects for dynamic controlled atmosphere (DCA) storage of fruits and vegetables, using chlorophyll fluorescence. http://www.harvestwatch.net/history.html. Accessed December 2014.

Rahman, A.S.A., D. Huber, and J.K. Brecht. 1993. Respiratory activity and mitochondrial oxidative capacity of bell pepper fruit following storage under low O_2 atmosphere. *Journal of the American Society for Horticultural Science* 118: 470–475.

Retamales, J., and R. Campos. 1997. Extremely low ethylene levels in ambient air are still critical for kiwifruit storage. *Acta Horticulturae* 444: 573–578.

Retamales, J., R. Campos, and D. Castro. 1998. Ethylene control and ripening in Packham's Triumph and Beurre Bosc pears. *Acta Horticulturae* 475: 559–566.

Rogiers, S.Y., and N.R. Knowles. 2000. Efficacy of low O_2 and high CO_2 atmospheres in maintaining the postharvest quality of saskatoon fruit (*Amelanchier alnifolia* Nutt.). *Canadian Journal of Plant Science* 80: 623–630.

Romero, I., M.T. Sanchez-Ballesta, R. Maldonado, M.I. Escribano, and C. Merodio. 2008. Anthocyanin, antioxidant activity and stress-induced gene expression in high CO_2-treated table grapes stored at low temperature. *Journal of Plant Physiology* 165: 522–530.

Satyan, S., K.J. Scott, and D. Graham. 1992. Storage of banana bunches in sealed polyethylene tubes. *Journal of Horticultural Science* 67: 283–287.

Schotsmans, W., A. Molan, and B. MacKay. 2007. Controlled atmosphere storage of Rabbiteye blueberries enhances postharvest quality aspects. *Postharvest Biology and Technology* 44: 277–285.

Schouten S.P., Prange R.K., and T.R. Lammers. 1997. Quality aspects of apples and cabbages during anoxia. *Proceedings of the 7th international controlled atmosphere research conference* Davis CA. USA, vol. 2, 189–192.

Scott, K.J., J.R. Blake, G. Strachan, B.L. Tugwell, and W.B. McGlasson. 1971. Transport of bananas at ambient temperatures using polyethylene bags. *Tropical Agriculture Trinidad* 48: 245–253.

Sharples, R.O. 1989. Kidd, F., and West, C. In *Classical papers in horticultural science*, ed. J. Janick, 213–219. New Jersey: Prentice Hall.

Stow, J.R., C.J. Dover, and P.M. Genge. 2000. Control of ethylene biosynthesis and softening in 'Cox's Orange Pippin' apples during low-ethylene, low-oxygen storage. *Postharvest Biology and Technology* 18: 215–225.

Strop, I. 1992. *Effects of plastic film wraps on the marketable life of asparagus and broccoli.* MSc thesis, Silsoe College, Cranfield Institute of Technology, UK.

Terry, L.A., T. Ilkenhans, S. Poulston, L. Rowsell, and A.W.J. Smith. 2007. Development of new palladium-promoted ethylene scavenger. *Postharvest Biology and Technology* 45: 214–220.

Thompson, A.K., 2010. *Controlled atmosphere storage of fruits and vegetables.* Second edition. CAB International, Oxford, UK.

Van Der Sluis, A.A., M. Dekker, A. De Jager, and W.M.F. Jongen. 2001. Activity and concentration of polyphenolic antioxidants in apple: Effect of cultivar, harvest year, and storage conditions. *Journal of Agricultural and Food Chemistry* 49: 3606–3613.

Villatoro, C., G. Echeverria, J. Graell, M.L. Lopez, and I. Lara. 2008. Long-term storage of pink lady apples modifies volatile-involved enzyme activities: Consequences on production of volatile esters. *Journal of Agricultural and Food Chemistry* 56: 9166–9174.

Wang, C.Y. 1990. Physiological and biochemical effects of controlled atmosphere on fruit and vegetables. In *Food preservation by modified atmospheres*, eds. Calderon, M., and R. Barkai-Golan, 197–223. Boca Raton, Ann Arbor, Boston, USA: CRC Press.

Wardlaw, C.W. 1938. Tropical fruits and vegetables: An account of their storage and transport *Low Temperature Research Station, Trinidad Memoir* 7, Reprinted from *Tropical Agriculture Trinidad* 14.

Wills, R.B.H., S. Pitakserikul, and K.J. Scott. 1982. Effects of pre-storage in low O_2 or high CO_2 concentrations on delaying the ripening of bananas. *Australian Journal of Agricultural Research* 33: 1029–1036.

Wojciechowski, J. 1989. Ethylene removal from gases by means of catalytic combustion. *Acta Horticulturae* 258: 131–142.

Wollin, A.S., Little, C.R., and J.S. Packer. 1985. Dynamic control of storage atmospheres. In: *Controlled atmospheres for storage and transport of perishable agricultural commodities*, ed. S. Blankenship, 308–315. Proceedings of the 4th National Controlled Atmosphere Research Conference, Raleigh, North Carolina, USA.

Wright, A.H., J.M. DeLong, A.H.L.A.N. Gunawardena, and R.K. Prange. 2012. Dynamic controlled atmosphere (DCA): Does fluorescence reflect physiology in storage? *Postharvest Biology and Technology* 64: 19–30.

Zheng, Y.H., C.Y. Wang, S.Y. Wang, and W. Zheng. 2003. Effect of high-oxygen atmospheres on blueberry phenolics, anthocyanins, and antioxidant capacity. *Journal of Agricultural and Food Chemistry* 51: 7162–7169.

Chapter 3
Hypobaric Storage

Introduction

The partial pressure of each component gas in air is changed in direct proportion to the total pressure as a consequence of the property of gases as described in Dalton's Law. John Dalton was a British chemist and natural philosopher who was born in 1766 and published Dalton's Law in 1801, although the French chemist and physicist Louis Joseph Gay-Lussac claimed that the "Law" had previously been described during the French Revolution and attributed it to Citizen Charles. Dalton's Law states that the volume of a gas maintained under constant pressure increases for equal increments of temperature by a constant fraction of its original volume and this fraction is the same whatever is the nature of the gas. For example where the pressure remains constant a mass of gas, whose volume is 1,000 at 0 °C, becomes 1,366.5 at 100 °C. The normal atmospheric pressure at sea level is 101.32494 kPa where the O_2 is equivalent to about 21 %. Therefore when the absolute total pressure is reduced to 10.13494 kPa, the O_2 partial pressure is equivalent to about 2.1 % O_2 (volume:volume) at atmospheric pressure. As Dalton's Law states the same is true for all other gases. In terms of its application to postharvest horticulture, the reduction in pressure reduces the partial pressure of O_2 and thus its availability to the flower, plant, fruit or vegetable in the store as Dalton's Law indicates the reduction in the partial pressure of the O_2 is proportional to the reduction in pressure. However, in the storage of fruit, vegetables and flowers the humidity must be kept high and this water vapour in the store atmosphere has to be taken into account when calculating the partial pressure of O_2 in the store. To do this, the humidity must be measured and the vapour pressure deficit can be calculated from a psychometric chart. This is then included in the following equation:

$$\frac{P_1 - \text{VPD} \times 21}{P_0} = \text{partial pressure of oxygen in the store}$$

© The Author(s) 2016
A.K. Thompson, *Fruit and Vegetable Storage*, SpringerBriefs in Food, Health, and Nutrition, DOI 10.1007/978-3-319-23591-2_3

where:

P_0 outside pressure at normal temperature (kPa)
P_1 pressure inside the store (kPa)
VPD   vapour pressure deficit inside the store (kPa)

Hypobaric storage has also been referred to as low pressure storage, LPS and sub-atmospheric pressure storage. Exposure of human beings and other animals to hypobaric conditions can result in hypoxia, which has been a well-recognised human illness for centuries and has been the subject of considerable research, for example Joanny et al. (2001) and Dolt et al. (2007). Hypobaric chambers are used for research and training on the effects of low O_2 and low atmospheric pressure on pilots and recent research is being carried out with a major aircraft manufacturer on the effect of aircraft cabin altitude on passenger discomfort, especially in long flights. Burg (2004) reviewed the considerable literature on the effects of hypobaric conditions on the storage and transport of meat, as well as horticultural produce. An example of meat transport was given by Sharp (1985) who reported on the successful transport of lamb carcasses at 0.006 atmosphere ($\approx$4.6 mm Hg) in a hypobaric container with an inner wall temperatures of about -1.5 °C and showed a mean weight loss of 2.5 % during 40 days storage. He emphasised the importance of precooling the carcasses before loading.

In relation to postharvest of fruit and vegetables different publications refer to the pressure data in different measurements. These include millimetres of mercury or mm Hg where atmospheric pressure at sea level is 760 mm Hg. This has been commonly used, but this is not an SI unit. The SI unit is a Pascal. Blaise Pascal was a French philosopher and mathematician, who was born in 1623 and died in 1662 and around 1647 he made discoveries concerning the weight of the atmosphere. A Pascal is defined as 1 Pascal $=$ 1 Newton m^2. A Newton is the SI unit of force, which acting on 1 kg of mass increases its velocity by 1 m s^{-1} every second along the direction it acts. Isaac Newton was a British mathematician who was born in 1642 and died in 1727. So 101.32494 kPa $=$ 760 mmHg $=$ 1 atmosphere. Pressure can be referred to as a standard atmosphere (1 atmosphere abbreviated as 1 atm) and reduced pressures as a fraction of atmosphere or of normal pressure e.g. $^1/_3$ atmosphere $=$ 253 mm Hg $=$ 33.775 Pascals. SI units use pressure measurements in Pascals where 1 mm Hg $=$ 133.322368 Pascals and kilopascals (kPa) is most commonly used in both hypobaric storage and controlled atmosphere storage where 1 mm Hg $=$ 0.133322368 kilo Pascals (kPa). Mega Pascals are also used, particularly for hyperbaric storage, and 1 MPa $=$ 1,000 kPa. 1 atmosphere $=$ 0.101325 MPa. Other terms that are used are bar where 1 mm Hg $=$ 0.00133322368 bar and Torr where 1 mm Hg $=$ 1 Torr and 1 Torr $=$ 133.322368 Pascals, 1 megapascals $=$ 1,000 kilopascals

(kPa) = 7,600 mm Hg, 1 megapascals = 7500.615613 mm Hg. The following is
given for ease of conversion when reading the literature:

mm mercury = kilo Pascals (kPa)	kilo Pascals = mm mercury (mm Hg)
1 mm Hg = 0.133322368	1 kPa = 7.6
10 mm Hg = 1.33	5 kPa = 38
20 mm Hg = 2.66	10 kPa = 76
24 mm Hg = 3.2	50 kPa = 380
25 mm Hg = 3.33	100 kPa = 760
40 mm Hg = 5.33	
50 mm Hg = 6.67	
60 mm Hg = 8 60	
80 mm Hg = 10.67	
100 mm Hg = 13.3	
500 mm Hg = 66.67	
760 mm Hg = 101.325	

Considerable work has been reported on the effects of hypobaric condition on
the human body, which can cause hypobaric hypoxia. This is a condition where
the body is deprived of a sufficient supply of O_2 for normal tissue metabolism and
affects the body's ability to transfer O_2 from the lungs to the bloodstream. It is
usually associated with exertion at higher altitude and can lead to a loss of cog-
nition. Considerable work has been published on hypobaric hypoxia especially
because of its implication to aerospace technology and mountain climbing.

History

Application of reduced pressure in postharvest agriculture has been used for
many decades. For example, Back and Cotton (1925) experimented with exposure
of horticultural crops to a vacuum for pest control and Bare (1948) showed that
reduced pressure reduced insect infestation in stored tobacco leaves. However,
perhaps the first records of hypobaric conditions being applied to the storage of
fruit and vegetables was by Workman et al. (1957) who found the respiration
rate of tomatoes was reduced when they were stored at 20 °C under 88 mm Hg
compared to those stored at 20 °C under atmospheric pressure. At the same time
Stoddard and Hummel (1957) stored several types of fruits and vegetables in
household refrigerators and found that those stored under 658–709 mm Hg had
increased postharvest lives of 20 to 92 %, depending on the crop, compared to
those stored in the same refrigerators at atmospheric pressure. Burg and Burg
(1965) and Burg and Burg (1966b) reported effects of hypobaric storage on fruits,
vegetables and flowers and Burg and Burg (1966a) showed delays in ripening of
bananas when they were stored under 125–360 mm Hg compared to those under
atmospheric pressure. They also showed that 50 % of limes stored at 15 °C took
about 10 days for 50 % of the green coloured fruit to turn yellow, while this was

delayed to 56 days in storage under 152 mm Hg at the same temperature. Burg (1975) reported that Lula avocados remained firm for 3.5 months under hypobaric conditions and then ripened normally upon removal. This length of storage is almost twice that reported Lula avocados held under normal atmospheric pressure controlled atmosphere storage (Hatton and Reeder 1965). Burg (1967) took out a patent in the USA for the application of hypobaric storage and he gave a detailed evaluation and justification of the award of the patent in Burg (2004). Stanley P. Burg was listed as the inventor and Grumman Allied Industries Incorporated as the original assignee. In support of the application several examples of the effects on horticultural products, meat and fish were given including those in Table 3.1.

Another patent for a hypobaric storage device, whose inventors and applicants were Peter Carlson and Lawrence P Kunstadt, had a publication date of 2001. Carlson's and Kunstadt's device was described in detail including "allows for the extended storage of all types of oxidizable materials, including foodstuffs such as fruits and vegetables; and inorganic materials such as oxidizable metals". The stored items are preserved solely through a reduction in air pressure, without a reduction of temperature, or the active addition and disposal of gases. This reduced pressure storage allows stored foodstuffs to preserve their original taste without hardening, and organic non-foodstuffs to be preserved without water condensation. Burg's patent describes succinctly hypobaric storage and is quoted as "The preservation of metabolically active matter such as fruit, vegetables, meat, fowl, shrimp, fish, other food, cut flowers, cuttings, foliage plants and the like is disclosed, characterised by storage at controlled and correlated conditions of hypobaric pressure, temperature, humidity, air circulation and air exchange. A non-deleterious gas such as air is humidified by contacting it with heated water from a supply, and then the humid air is passed through, and when advisable, recirculated and/or re-humidified within a storage chamber containing the metabolically active matter. The humidity is maintained within the range of about 80–100 % r.h. and the pressure is maintained continuously or intermittently at a selected value at least slightly higher than the vapour pressure of the water in the stored commodity". Burg (2004) refers to legal action in which he was involved related to "LP patents" and tax implications of those patents that were eventually resolved. Some information on patents related to storage of fruit and vegetables are given in Table 3.2.

In Britain, work on hypobaric storage was started at the East Malling Research Station in 1971 (Sharples 1971, 1974; Anonymous 1975). Langridge and Sharples (1972) developed apparatus where bushel-sized samples (about 36 L) of fruit could be stored under hypobaric conditions and compared with those stored at atmospheric pressure as well as fruits stored under controlled atmospheres of reduced O_2 and perhaps increased CO_2. Work on hypobaric storage in Britain was also undertaken at the Tropical Products Institute in London (Hughes et al. 1981) and at the National Vegetable Research Station at Wellesbourne (Ward 1975), but Colin Ward and Bill Tucker had "teething problems getting the equipment to work and the industry just weren't interested at the time" (David Gray personal communication). Considerable and very comprehensive work was also carried out in

Table 3.1 Effects of hypobaric storage on the postharvest life of some fruits and vegetables. Taken with modifications from Burg S P 1977 Patent US4061483A http://www.google.co.uk/patents/US4061483?utm_source=gb-gplus-sharePatent US4061483—low temperature hypobaric storage of metabolically active matter. Accessed September 2014

Crop	Temperature	Pressure	Storage time
Bartlett, clapp and comice pears	−1 to 1 °C	Atmospheric	1½ to 3 months
		60 mm Hg.	4 to 6 months
McIntosh, red delicious, golden delicious and Jonathan apples	−1 to 2 °C	Atmospheric	2 to 4 months
		60 mm Hg.	6 months
Waldin avocados	10 °C	Atmospheric	12 to 16 days
		60 to 80 mm Hg	30 days
Lula avocados	8 °C	Atmospheric	23 to 30 days
		40 and 80 mm Hg	75 to 100 days
Booth 8 avocados	8 °C	Atmospheric	8 to 12 days
		40 and 80 mm Hg	about 45 days
Fresh green onions (scallions)	0 to 3 °C	Atmospheric	2 to 3 days
		60 to 80 mm Hg	More than 3 weeks
Green peppers (capsicums)	8 to 13 °C	Atmospheric	16 to 18 days
		60 to 80 mm Hg	46 days
Snap beans	5 to 8 °C	Atmospheric	7 to 10 days
		60 mm Hg	26 days
Cucumbers	10 °C	Atmospheric	10 to 14 days
		80 mm Hg	49 days
Pole beans	8 °C	Atmospheric	10 to 13 days
		60 mm Hg	30 days
Tioga and Florida 90 strawberries	0 to 2 °C	Atmospheric	5 to 7 days
		80 to 200 mm Hg	4 to 5 weeks
Blueberries	0 to 1 °C	Atmospheric	4 weeks
		80 to 200 mm Hg	at least 6 weeks
Iceberg lettuce	0 to 4 °C	Atmospheric	2 weeks
		150 to 200 mm Hg	about 4 weeks
Ruby red grapefruit	6 °C	Atmospheric	4 to 6 weeks
		80–150 mm Hg	90 days
Mature green tomatoes	13 °C	Atmospheric	2 weeks
		80 mm Hg	8 weeks

Germany, for example Bangerth (1973, 1974, 1984 and 1987), in Spain (Alvarez 1980), in Canada (Lougheed et al. 1974, 1977) in Israel (Apelbaum and Barkai-Golan 1977; Apelbaum et al. 1977a, b and Aharoni et al. 1986), in China (Chang Yan Ping 2001 and Cao Zhi Min 2005) and in Italy (Romanazzi et al. 2001, 2008). However, most of the researches on hypobaric storage and its commercial application have been carried out in the USA particularly by Stanley Burg and many other workers including David Dilley.

Table 3.2 Patents related to hypobaric storage of fruits and vegetables as presented in http://www.google.com.ni/patents/US3333967. Accessed September 2014

Citing patent	Filing date	Publication date	Applicant	Title
US3913661[a]	26 March 1974	21 October 1975	Grumman allied industries	Low pressure storage of metabolically active material with open cycle refrigeration
US4331693[a]	8 September 1980	25 May 1982	Polska Akademia Nauk, Instytut Katalizy I Fizykochemii Powierzchni	Method for storage of horticultural products in freshness
US4506599[a]	20 December 1983	26 March 1985	Polska Akademia Nauk, Instytut Katalizy I Fizykochemii Powierzchni	Device for removal of ethylene from fruit storage chambers
US4655048[a]	6 December 1985	7 April 1987	Burg Stanley P	Hypobaric storage of non-respiring animal matter without supplementary humidification
US4792455[a]	10 November 1986	20 December 1988	Ottmar Tallafus	Method for preserving fruits and vegetables
US4857350[a]	10 March 1987	15 August 1989	Kiyomoto Tekko Kabushiki Kaisha	Method for maintaining or restoring freshness of plants, vegetables or fruits by treating with water supersaturated with air
US4884500[a]	13 March 1989	5 December 1989	Yoshihiko Iwasaki	Apparatus for maintaining or restoring freshness of vegetable body
US6165529[a]	14 December 1999	26 December 2000	Planet Polymer Technologies, Inc.	Coating exterior surface of produce with coating composition comprising aqueous solution of 1-20 % by weight of hydrolyzed cold water insoluble polyvinyl alcohol, 0.1–10 % cold water soluble starch

(continued)

Table 3.2 (continued)

Citing patent	Filing date	Publication date	Applicant	Title
US6203833	16 February 2000	20 March 2001	Planet Polymer Technologies, Inc.	Process for preserving fresh produce
US7650835	14 Dec 2004	26 January 2010	Russ Stein	Produce ripening system
US8632737	15 April 2010	21 January 2014	Atlas Bimetals Labs, Inc.	Systems and methods for controlled pervaporation in horticultural cellular tissue
US20130295247[a]	8 July 2013	7 November 2013	Hussmann Corporation	Table with ethylene scrubber
DE2422983A1[a]	13 May 1974	27 November 1975	Grumman Allied Industries	Verfahren und vorrichtung zum klimatisieren einer lagerkammer
WO1988008106A1[a]	6 April 1987	20 October 1988	Stanley R Burg	Hypobaric storage of non-respiring animal matter without supplementary humidification
WO2001000074A1[a]	28 June 2000	4 January 2001	Carlson Peter	Hypobaric storage device

[a]Cited by examiner

Besides its high cost, Burg (2014) evaluates the other reasons why hypobaric technology has had only limited impact on the food industry. His contentions include:

problems of leakage in laboratory apparatus used for testing hypobaric conditions, experimental error caused by humidifying air at 1 atmosphere pressure instead of at the low storage pressure,
insufficient air changes so that the stored commodity consumed all available O_2 and suffered anaerobic damage
laboratory apparatus installed in cold rooms that had a non-uniform air distribution pattern that produced a "cold spot" on the vacuum chamber's surface. This created an evaporation/condensation cycle between the commodity and chamber cold spot.
the commodity was stored in a sealed system that accumulated an active ethylene concentration between occasional ventings.

Mode of Action of Hypobaric Conditions

The literature reports several effects of hypobaric conditions on fresh fruits and vegetables that can lead to extension in the postharvest life. It has been well established that the reduced O_2 in controlled atmosphere storage is a major factor in extending their postharvest life. However, in addition to the effects of reduced O_2 on their respiration rate, hypobaric conditions have additional effects. Since hypobaric chambers are constantly ventilated and air removed ethylene produced by the fruit or vegetable will be constantly removed. Ethylene has also been shown to be removed more quickly from the plant cells where it is being synthesised. This ethylene removal would therefore reduce its effect in the fruit ripening process and also reduce the effects on the development of some physiological disorders associated with ethylene accumulation. Huiyun Chen et al. (2013a) also reported that freshly bamboo shoots stored at 2 °C under 50 kPa for 35 days had reduced ethylene production, compared to those stored under atmospheric pressure, which delayed their softening. Burg (2004) reviewed several publications that reported inhibition of growth and sporulation of pathogenic fungi under hypobaric conditions that then resumed growth and sporulation when they were removed to atmospheric pressure, which may also be related to reduced O_2. There is also strong evidence that hypobaric conditions can help control insect infestation if fruits. For example, Davenport et al. (2006) reported that all the eggs and larvae of Caribbean fruit fly (*Anastrepha suspensa*) in mangoes stored at 13 °C under 15–20 mm Hg with ≥98 % r.h. were killed within 11 days, while a substantial number of eggs survived for the 14 days storage at 13 °C with ≥98 % r.h. under atmospheric pressure. Hypobaric conditions may have other effects on fruit physiology. For example, Jinhua Wang et al. (2015) concluded that the shelf life extension of honey peach (*Prunus persica*) in hypobaric storage could be due to their increased energy status, enhanced antioxidant ability and less membrane damage.

Technology

Hypobaric storage is a system of storing commodities while ventilating with air at less than atmospheric pressure. Since the crop in the hypobaric store is constantly respiring, it is essential that the store atmosphere is constantly being changed in order to maintain the desired O_2 level. This is achieved by a vacuum pump evacuating the air from the store. The store atmosphere is constantly being replenished from the outside. The air inlet and the air evacuation from the store are balanced in such a way as to achieve the required reduced pressure within the store. There are two important considerations in developing and applying this technology to crop storage. The first is that the store needs to be designed to withstand low pressures without imploding. The second is that the reduced pressure inside the store can result in rapid water loss from the crop. To overcome the first, stores have to be strongly constructed, for example with thick steel plate with a curved interior. For the second, the air being introduced into the store must be saturated (100 % r.h.) or as close to saturation as possible. If it is less than this, serious dehydration of the crop can occur.

The control of the O_2 level in the store can be very accurately and easily achieved and simply measured by measuring the pressure inside the store with a vacuum gauge. Hypobaric storage also has the advantage of constantly removing ethylene gas from the store which prevents it building up to levels which could be detrimental to the crop. In fact, Salunkhe and Wu (1975) commented that hypobaric storage "indiscriminately lowers the internal equilibrium content of all volatiles, including ethylene" of stored fruit and vegetables. This would have the effect of reducing the detrimental effects of ethylene on their postharvest life.

Various systems have been developed to test hypobaric storage but most are similar to the one described by Dilley (1977) as "The product is held in a vacuum tight compartment while being continuously ventilated with water-saturated air at absolute pressures ranging from 10 to 80 mm Hg". The equipment used by Wang and Dilley (2000) was 1,270 L vacuum vessels with a ventilation rate of one vessel void volume change per hour at 97 % r.h. The vessels were kept in a temperature controlled room. This type of system was also used by Hughes et al. (1981) where specially constructed 100 L steel barrels were used that had 4 cm thick transparent Perspex lids through which the produce could be observed during storage. Air was introduced through an inlet that had been bubbled through water in order to saturate it (although no measurement of the relative humidity was made) and a vacuum pump and vacuum gauge were attached to the outlet and the flow rate adjusted on the pump to give a flow rate of 5 L h^{-1}. The systems were kept in temperature controlled cabinets that were adjusted to $\pm$ 1 °C. Romanazzi et al. (2001) used a vacuum pump in 64 L gas-proof tanks in which the fruit were placed in these tanks at atmospheric pressure as controls or a vacuum applied with a vacuum pump and measured with an external vacuum meter. Al-Qurashi et al. (2005) used quart sized pressure cookers that were continuously evacuated by a belt-drive pump. The inlet air was humidified by bubbling it through water in 5 gallon containers then through a filter to prevent water from getting into the pressure cookers from the ventilated air. The air filter was ¼ filled with water and cellulose pads were inserted to increase the humidity of the air flowing to the pressure cookers. The pressure cookers were also sealed at the lids with 'Play Dough', allowing an airtight seal. Valve regulators, located between the filter and the pressure cookers, were used to maintain the desired pressures by admitting air at the proper rate. The pressure within the pressure cookers was monitored with pressure gauges placed at the top of the pressure cookers. Jiao et al. (2012a, b) used aluminium chambers ($0.61 \times 0.43 \times 0.58$ m) with a two-stage rotary vacuum pump regulated by a compact proportional solenoid valve controlled by a proportional/integral/derivative computer control system. Chamber pressure was monitored with a digital pressure gauge. A rotameter was used to adjust the air exchange rate and the ingoing rarefied air was passed through a humidifier before entering

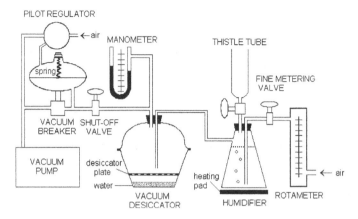

Fig. 3.1 Laboratory apparatus for experiments on hypobaric storage. *Source* Burg (2004, p. 366). Reproduced with permission of Dr Stanley Burg and the Commonwealth Agricultural Bureau International

the hypobaric chamber in order to keep the humidity near saturation. The relative humidity was calculated by measuring wet-bulb and dry-bulb temperatures having relatively high accuracy ($\pm$ 0.1 %). They experimented with various modifications and showed that added foam covering the chambers maintained the temperature of the inside air to within $\pm$ 0.1 °C. The regulating system kept pressure to within $\pm$ 1 % of the set point and maintained humidity at >98 % r.h. under various air exchange rates and pressures (the measurements inside the cabinet varied between 98.40 and 99.35 r.h.), with a chamber leakage rate of 0.009 kPa h^{-1} and hypobaric system leakage rate of 0.48 kPa h^{-1}. Spalding and Reeder (1976) used pure CO_2 and O_2 from gas cylinders that were metred into a 150 ml glass mixing chamber and their flow rates regulated to supply the required mixture of O_2 and CO_2. Flow into the chamber was controlled at 110 ml min^{-1} (1 air change h^{-1}). Chamber pressures were maintained at about 2 mm Hg and the humidity at 98–100 % r.h. as determined by a humidity-sensing element. Jamieson (1980) used vacuum desiccators for the fruit or vegetables to be tested. A vacuum pump sucked air from the desiccators and air was allowed in by first passing it through a flask containing water that was constantly heated on an electric plate to maintain near saturation humidity. Air flow and pressure inside the desiccators were measured and controlled with a flow meter, a vacuum gauge, a pressure regulator and a needle valve. Burg (2004) described equipment used in laboratory scale hypobaric experiments (Fig. 3.1).

As has been mentioned above hypobaric conditions have been used to simulate at least one effect of controlled atmosphere storage, that is the reduced O_2 supply. The technology used in controlled atmosphere storage has developed over the years since it was first introduced, but the principal remains the same. Humidity control is applied in controlled atmosphere storage and this is more important in hypobaric storage but control of temperature is equally important in both technologies.

Jiao et al. (2012a, b) described a hypobaric chamber called VivaFresh that had been made in 2007 by Atlas Technologies, Incorporated, Port Townsend, Washington, USA. VivaFresh are aluminium chambers (0.61 m long × 0.43 m wide × 0.58 m high) with a two-stage rotary vacuum pump regulated by a compact proportional solenoid valve controlled by a proportional/integral/derivative computer control system. Chamber pressure was monitored with a digital pressure gauge and a rotameter was used to adjust the air exchange rate and the ingoing rarefied air was passed through a humidifier before entering the hypobaric chamber in order to keep the humidity near saturation. The temperature inside the chamber and the exterior chamber wall and humidity was calculated by measuring wet-bulb and dry-bulb temperatures using calibrated YSI 55000 Series GEM thermistors with a relatively high accuracy of ± 0.1 % (Wang et al. 2003). Temperature variation of the chamber wall was controlled to within ± 0.2 °C and the inside air to within ± 0.1 °C. Humidity measured inside the chambers varied between 98.4 and 99.35 % r.h. The pressure was within 1 % of the set point and O_2 concentration could be controlled at <0.6 % when the pressure was less than 3.3 kPa. The leakage rate of the chamber was 0.01 kPa h^{-1}.

Burg (2014) commented that the academic belief that "hypobaric storage is a flawed technology originated from experimental errors in low pressure research caused by non-precise temperature control, cold spots on the vacuum chamber's surface, humidifying at atmospheric pressure rather than a low pressure, inadequate air changes, leaky vacuum chambers and a failure to realise that the high turgor pressure of plant cells prevents low pressure storage from causing volatiles to boil and outgas (release of a gas that was dissolved, trapped or absorbed)". He further commented that "…experimental errors by academics and other concerns have prevented hypobaric storage from achieving more widespread adoption".

Transport

In the 1970s the Grumman Corporation in the USA developed and constructed a hypobaric container which they called 'Dormavac'. It was operated at 2.2–2.8 °C and a pressure of 15 mm Hg and they tested it in commercial situations, but were unable to make it profitable, resulting in eventual losses of some $50 million (Anonymous undated). Burg (2014) reported that between 1976 and 1982 prototype Grumman Dormavac hypobaric intermodal containers successfully exported asparagus, mangoes, papayas and fresh meat. Grumman Corp. and Armour & Co. were awarded the US Food Technology Industrial Achievement award for developing hypobaric transportation and storage systems. Burg (2004) described the Dormavac system in detail with a general outline given in Fig. 3.2.

Alvarez (1980) described experiments where papaya fruits were subjected to sub-atmospheric pressure of 20 mm Hg at 10 °C and 90–98 % r.h. for 18–21 days

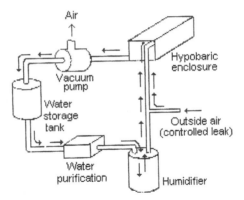

Fig. 3.2 Grummen/Dormavac vacuum/humidity subsystem uses a two-stage vacuum pump to reduce pressure, change the air in the hypobaric enclosure and remove commodity generated gases and water vapour. *Source* Burg (2004, p. 500). Reproduced with permission of Dr Stanley Burg and Food Technology

during shipment in a hypobaric container from Hawaii to Los Angeles and New York. Both ripening and disease development was inhibited. Fruits ripened normally after removal from the hypobaric containers, but abnormal softening unrelated to disease occurred in 4 to 45 % of fruits of one packer. It was found that hypobaric stored fruits had 63 % less peduncle infection, 55 % less stem-end rot and 45 % fewer fruit surface lesions than those stored in a refrigerated container at normal atmospheric pressure.

A more recent system of hypobaric storage, which controls water loss from produce without humidifying the inlet air with heated water, was designed by Stanley Burg (Burg 1993). The system was described by Burg (2004) and is called VacuFresh[sm] (Fig. 3.3). Humidity control is achieved by slowing the evacuation rate of air from the storage chamber to a level where water evaporated from the produce by respiratory heat exceeds the amount of water required to saturate the incoming air. Using this technique with roses stored at 2 °C and 3.33 × 103 Pa, Burg (1993) found that flowers stored for 21 days with or without humidification at a flow rate of 80–160 cm^3 min^{-1} lost no significant vase life compared with fresh flowers. It had a very slow removal of air and therefore it was claimed that there was no desiccation problem of the fruit, flowers or vegetables transported in these type of container.

VacuFresh Corporation is a South Africa company (Welfit Oddy (Pty.) Limited) that produce hypobaric intermodal shipping containers, which are claimed to be energy efficient (Fig. 3.4) and capable of controlling O_2 levels very precisely. This refrigerated tank container cool the cargo by circulating brine or synthetic oil around external cooling coils in the tank. The system is primarily used to transport liquid chemicals, pharmaceuticals, food products and beverages.

Davenport et al. (2006) tested VacuFresh containers on mature mangoes, which they reported had previously been shown to extend their storage life far longer

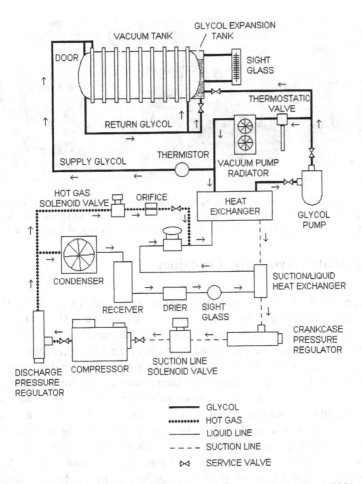

Fig. 3.3 General schematic of VacuFresh[sm] refrigeration system. *Source* Burg (2004, p. 508). Reproduced with permission of Dr Stanley Burg and the Commonwealth Agricultural Bureau International

than was possible using other technologies. It was claimed that the intermodal container provided a hypobaric atmosphere that can store mangoes for up to two months with no deterioration of quality even of fully mature fruit. They tested the containers as a possible system of insect control to comply with phytosanitary regulations during transport of fresh produce to certain destinations. The ability of Caribbean fruit fly (*Anastrepha suspensa*) eggs and larvae to survive simulated optimal hypobaric conditions of 13 °C with 15 and 20 mm Hg with $\geq$98 % r.h. was tested. Nearly 98 % of the eggs and larvae were killed within 1 week and all eggs were killed by 11 days exposure whereas a substantial number of eggs survived for the 14 days of the trial at 13 °C with $\geq$98 % r.H. under atmospheric pressure.

Fig. 3.4 Intermodal tank container from Welfit Oddy (Pty) Ltd. Kurland Road, Perseverance, Port Elizabeth 6001, South Africa. This particular container was sold to Klinge Corporation of Pennsylvania in USA, Egypt and Denmark. Reproduced with permission of Tim McLaren

Non-ventilated Hypobaric Containers

A different hypobaric system was described by Knee and Aggarwal (2000) who used plastic containers, capable of being evacuated to 380 mm Hg with a vacuum pump. They found that this required 18 strokes of the pump for a container with a nominal volume of 500 ml capacity and 24 strokes for a 750 ml capacity container. These were placed in a refrigerators run at either 4 or 8 °C depending on the product to be stored. Overall the vacuum containers showed little advantage over conventional plastic containers for the types of produce tested. The multinational company VacuFresh also make small plastic containers of 1.2–2.8 L capacity that are sold together with a hand pump. They are used for storing fruits and vegetables, etc. under hypobaric conditions in a domestic refrigerator. The company claim in their advertisements "Keeps food fresh up to 5 times longer. Vacuum locks in freshness and prevents premature spoilage and food decay". In the mid 1970s Prodesarrollo in Colombia (A.K. Thompson unpublished) investigated the possibility of storing vegetables (potatoes, carrots and cabbages) in a brick built store room at a cafe/ski centre at about 4800 m that would have a barometric pressure of about 57 kPa (430 mm Hg) where the temperature was also low. The trial was on the slopes of the volcano Nevado del Ruiz near the town of Manizales in the Departamento Caldas whose peak is 5321 m above sea level. The trial was unsuccessful due to rapid desiccation of the vegetables, and because of transport problems it was not repeated.

Low Oxygen

Reducing the O_2 levels in fruits and vegetables stores is known to increase their postharvest life and there are various ways of achieving this including controlled atmosphere storage and hypobaric storage. A major effect of hypobaric conditions is the reduction in the partial pressure of O_2. Controlled atmosphere storage has been used commercially for almost 100 years and is increasing in its commercial application. Generally, crops stored under controlled atmospheres of reduced O_2 and, to a lesser extent, increased CO_2 have a longer storage life because the rate of the metabolic processes is slower. Particularly with climacteric fruit this would slow ripening and deterioration so that when they have been stored for protracted periods they may well be less ripe than fruits stored in air at the same temperature. The effects of reduced O_2 levels on postharvest responses of fruits and vegetables were reviewed and summarised by Thompson (2010) as follows:

- altered texture
- changed fatty acid synthesis
- delayed breakdown of chlorophyll
- delayed ripening of climacteric fruit
- development of fungal and bacterial diseases
- development of physiological disorders
- formation of flavour and odours
- prolonged storage life
- reduced degradation rate of soluble pectins
- reduced rate of production of ethylene
- reduced respiration rate
- reduced substrate oxidation
- survival of pests

Effects

The effects of hypobaric storage on fruits and vegetables have been reviewed by Salunkhe and Wu (1975), Burg (1975, 2004, 2014). The reviews showed considerable extension in the storage life of a wide range of crops when reduced pressure was combined with refrigeration compared to refrigeration alone. Specific beneficial effects of hypobaric storage have been reported for various fruits and vegetables as well as other foods and flowers. For example, under continuously ventilated partial pressure, CO_2, ethylene and various volatile by-products of metabolism rapidly diffuse out of the crop and are flushed from the storage chamber. As a consequence of the low partial pressure and the low levels of ethylene in the atmosphere, ripening and senescence of fresh fruits and vegetables are delayed and their storage life is extended. Other publications describing positive effects of hypobaric storage include: Davenport et al. (2006), Knee and Aggarwal (2000) and Li et al. (2008) and Tolle (1969) concluded that the chief merits of hypobaric storage were the continuous removal of ethylene from the storage environment and the lowering of the partial pressure of O_2.

Considerable work has shown that low O_2 in the storage atmosphere can slow the respiration rate of fruits and vegetables, for example by Kidd and West (1927). Also Choudhury (1939) found that the respiration rate of fruit and vegetables increased with increasing in O_2 concentration when they tested them over the range of 6.2–98.6 kPa. These effects could result in the slowing of ripening of climacteric fruits such as tomatoes and reduced yellowing of vegetables such as broccoli and bean leaves (Nilsen and Hodges 1983). However, Paul and Ferl (2006) working on the implications of hypobaric conditions for space exploration, studied the effects of hypobaric stress on Thale Cress plants (*Arabidopsis thaliana*). Less than half of their genes induced under hypobaric conditions were induced by hypoxia, establishing that the response to hypobaric conditions was more complex than just an adaptation to low partial pressures of O_2. They also reported that the genes of *A. thaliana* induced by hypobaric conditions confirmed that water movement was a paramount issue in plants and that even small changes in atmospheric pressure had biological consequences. Overall the effects of hypobaric conditions on the postharvest life of fruit and vegetables can probably be accounted for by several interacting factors.

Diffusion

Hypobaric conditions have been shown to accelerate the outward diffusion of gases from the internal tissues of horticultural crops during storage (Goszczynska and Ryszard 1988). Burg and Kosson (1983) reported that hypobaric storage lowers the internal equilibrium content of volatiles, including ethylene.

Ethylene

Removal of ethylene from the internal atmosphere of the fruit or vegetable and flushing it from the store can occur under hypobaric conditions. The ethylene concentration within a mature unripe apple was given as about 0.1 parts per million by Ryall and Pentzer (1974) while at 76 mm Hg this level would be reduced to 0.01 parts per million, a level that they report is insufficient to initiate ripening. Sharples and Langridge (1973) found that during storage of the apple cultivar Cox's Orange Pippin in thick walled steel drums at 3.3 °C with an airflow of 15 L h^{-1} the ethylene level was 5 parts per million in the store with 0.5 atmosphere compared to 10 parts per million for those stored in atmospheric pressure. McKeown and Lougheed (1981) commented that hypobaric storage "provides a simple means for reducing the effect of ethylene-producing crops upon vegetables in the same storage, but with no ethylene source in the storage environment there seems to be little benefit of lowering the partial pressure of oxygen". Burg (2014) reported that "several researchers claimed that hypobaric storage cannot displace 'active-bound ethylene' from within plant tissues and only influences ripening or senescence by lowering O_2. This opinion was disproved by studies showing that ethylene's measured dissociation constant from its receptor site has the same value as the applied ethylene concentration causing a half-maximal biological response. 'Bound' and intercellular ethylene equilibrated within 15 min after the elongation growth of plants that do not produce autocatalytic ethylene was inhibited by applied ethylene and the plants were transferred to fresh atmospheric air".

Other Volatiles

Hypobaric storage could possibly remove other volatiles produced by fruits and vegetables during storage. Bangerth (1984) reported that apples stored for protracted periods under 50–75 mm Hg did not develop normal aroma and flavour when they were removed to atmospheric pressure and allowed to ripen. There is some indication that this may be due to the low partial pressure of O_2 since the same effect has been reported for apples in controlled atmosphere storage (Burg 2004). Wang and Dilley (2000) proposed that hypobaric ventilation removes a scald-related volatile substance that otherwise accumulates and partitions into the epicuticular wax of fruit stored under air at atmospheric pressure. They provided evidence that α-farnesene and 6-methyl-5-hepten-2-one accumulation in the epicuticular wax associated with hypobaric storage may be involved.

Oxygen

As indicated above a major effect of hypobaric conditions is to reduce the partial pressure of O_2. The effects of storage of fruit and vegetables in atmospheres containing low O_2 have been shown to have beneficial effects on their postharvest life. This was reviewed by Thompson (2010). Burg (2010) reported that "The high gaseous diffusion rate at a low pressure eliminates the commodity's surface to centre O_2 gradient created by respiratory O_2 consumption, causing different commodity types to have nearly identical low O_2 tolerances, near 0.1 %". Under hypobaric conditions, Burg (2004) reported that fermentation is not induced at O_2 levels as low as 0.06–0.15 %, while fermentation can occur under atmospheric pressure. Burton (1989) and Mapson and Burton (1962) reported that gaseous diffusion through the periderm of mature potato tubers was entirely or almost entirely through the lenticels. They reported that permeability of tubers ranged from about 0.7 to about 2.5 $mm^3\ cm^{-2}\ h^{-1}\ kPa^{-1}$, depending on maturity, time in storage and cultivar, and O_2 diffusion would occur with an O_2 deficit of about 0.4–0.5 %.

Carbon Dioxide

Burg (2004) observed that hypobaric conditions can decrease ambient and inter-cellular CO_2, which is an important advantage, providing benefits that were not duplicated by increasing CO_2 to levels given in some controlled atmosphere stor-age recommendations. However, Spalding and Reeder (1976) concluded that high CO_2 was necessary for the successful storage of Waldin avocados, even in hypobaric storage. Yahia (2011) commented that CO_2 was considered essential in controlling decay and ameliorating chilling injury in avocados and CO_2 could not be added in a low pressure system. Burg (2004) contended that the effects of hypobaric conditions on removing the CO_2, produced by respiration, from cells and intercellular spaces may result in reduced bacterial and fungal growth, better ascorbic acid retention of the fruit, inactivation of the ethylene forming enzyme and the prevention of succinate formation. In confirmation of the effect of CO_2 on bacterial growth, Wells (1974) reported that the postharvest pathogens, *Erwinia carotovora*, *E. atroseptica* and *Pseudomonas fluorescens* were unable to multiply in the very low CO_2 levels, as well as the low O_2 levels, which are to be found in the cells of vegetables stored under hypobaric conditions. However, Enfors and Molin (1980) found that when *Pseudomonas fragi* was grown at O_2 limita-tion (0.0025 atmosphere O_2) and exposed to 0.99 atmospheres CO_2, the inhibit-ing effect of the CO_2 was added to that of the O_2 limitation. They did not note any indications of a synergistic effect between CO_2 inhibition and O_2 limitation. Laurin et al. (2006) reported that cucumbers exposed to hypobaric conditions of 532 mm Hg for only 6 h may exhibit an indirect stress response that occurs only when the fruit were transferred to atmospheric pressure, preventing closure of sto-mata. They explained that this residual effect may have been due to the possibility that hypobaric conditions enhanced outward diffusion of CO_2, reducing intercel-lular CO_2 concentration and causing stomata to open. When the fruits were trans-ferred to atmospheric pressure stomata may still have remained open to restore the CO_2 concentration.

Respiration Rate

Hypobaric storage has been shown to reduce respiration rate compared to stor-age under atmospheric pressure. This has been shown on many fruits and vegeta-bles including sections of oat leaves (Veierskov and Kirk 1986) oranges (Min and Oogaki 1986), tomatoes (Workman et al. 1957), apples (Bubb 1975b), asparagus spears (Wenxiang et al. 2006; Li et al. 2006a) and cranberries (Lougheed et al. 1978). Cytochrome oxidase, the final electron-transferring enzyme of the res-piratory chain, has a great affinity to O_2 but, cytochrome oxidase is still able to operate under low O_2 pressure such as 0.01 atmosphere without its activity being altered (Mapson and Burton 1962; Burton 1989). However, Mapson and Burton

(1962) attributed the reduction of respiration rate, during hypobaric storage, to the malfunctioning of oxidases such as polyphenol oxidase or ascorbic acid oxidase. Hence, under the hypobaric condition of 0.70 atmosphere (532 mm Hg), it is expected that O_2 partial pressure is not sufficiently reduced to cause the respiration rate to decrease. In the tissues of higher plants, the effect of exposure to pure O_2 was to stop of carbon dioxide output, possibly the result of the inactivation of the associated enzyme systems (Caldwell 1965).

Chilling Injury

Hypobaric storage can reduce susceptibility to chilling injury. Chen et al. (2013a) reported that malondialdehyde (MDA) content in Chinese bayberry fruit (*Myrica rubra*) was related to chilling injury since it is considered to be an indicator of membrane lipid peroxidation caused by oxidative stress. Electrolyte leakages, as well as MDA content, are indicators of cell membrane damage. MDA content and electrolyte leakage are used to indicate lipid peroxidation of membrane lipids and membrane permeability, respectively, which increase during low temperature storage (Zhao et al. 2006). MDA content of the Chinese bayberry fruit under normal atmospheric pressure condition increased gradually during storage while storage for 15 days under hypobaric pressures of 85 ± 5, 55 ± 5 and 15 ± 5 kPa all inhibited the accumulation of MDA. Similar results had been obtained by Li et al. (2006) who reported that hypobaric storage could reduce MDA accumulation in asparagus.

Chlorosis

Nilsen and Hodges (1983) exposed bean leaves (*Phaseolus vulgaris*) to ethylene by dipping them in 30 parts per million Ethephon and storing them at 26 °C, which resulted in them becoming chlorotic more rapidly (reaching peak levels within 6 h) than those not treated. However, when the Ethephon treated leaves were stored under hypobaric conditions (200 millibars, with O_2 and CO_2 compositions set to approximate normal atmospheric partial pressures), chlorophyll loss was prevented.

Desiccation

Reduced pressure inside the store can result in rapid water loss from the crop since the boiling point of water reduces from 100 °C at atmospheric pressure to 0 °C at 4.6 mm Hg. Therefore, there is a clear tendency to desiccation under hypobaric storage and fruits and vegetables need to be retained at humidity as close as possible to saturation to limit weight loss. The particular hypobaric pressures used for fruits and vegetable storage can also affect their weight loss. Apelbaum et al. (1977b) tested the effect of hypobaric pressure storage on mango fruits and observed that at pressure below 50 mm Hg, mangoes underwent desiccation. Cicale and Jamieson (1978 quoted by Burg 2004) reported that their best results were storage of avocados at 61 mm Hg since lower storage pressure resulted in higher desiccation. Patterson and Melsted (1977) reported that cherries under hypobaric conditions resulted in some problems with desiccation especially at 41 mm Hg. Bubb (1975b) found that apples under hypobaric storage (35–40 and 70–80 mm Hg) had higher weight losses than the apples under atmospheric pressure or controlled atmosphere storage. Hughes et al. (1981) also reported increased desiccation of capsicums during hypobaric storage compared to those stored under atmospheric pressure. An et al. (2009) reported that curled lettuce had high moisture loss during storage under both 190 and 380 mm Hg compared with those under atmospheric pressure. Spalding and Reeder (1976) reported that average weight loss and shrivelling of limes were higher during storage under hypobaric conditions compared to those stored at atmospheric pressure. Conversely Burg (2004) found that at 0–3 °C spring onions, in general, lost less weight during hypobaric storage than during storage in air at atmospheric pressure. In general, the weight loss of radishes during storage at 1 °C under 56 mm Hg and near-saturation humidity was less than that which occurred during storage at atmospheric pressure (McKeown and Lougheed 1981). Cicale and Jamieson (1978 quoted by Burg 2004) found that avocados lost 1.2 % in weight at various pressures ranging from 61 to 203 mm Hg compared to 5.7 % under atmospheric pressure during storage at 6 °C for 35 days. Spalding and Reeder (1976) reported that humidity did not appear to be a factor in the storage life of avocados since the acceptability of avocados stored under hypobaric storage at 80–85 % r.h. and 98–100 % r.h. were not significantly different. All successful hypobaric systems have some form of humidification of the air as it enters the chambers, but Burg (2004) commented that in many cases these may have been insufficient. In data provided in the literature McKeown and Lougheed (1981) reported that the highest weight loss of asparagus was while the pressure was decreasing and humidity was not near saturation and they sprayed asparagus spears with water before hypobaric storage to reduce desiccation. Laurin et al. (2006) also reported that water can be sprayed on cucumbers to resolve the problem of insufficient relative humidity, causing desiccation during hypobaric storage. It was claimed that the VacuFresh system of hypobaric storage had a very slow removal of air and therefore there was no desiccation problem of the fruit and vegetables transported in these containers (Burg 2004).

   In Burg's hypobaric patent application he states "Though relative humidities of 80 % are usefully permissible, the preferable relative humidity of the air in the storage chamber should be higher than approximately 90 % for the storage of foodstuffs such as fruits and vegetables". (Burg 1976). Burg and Kosson (1983) commented that reduction in air pressure surrounding plant tissue will reduce the cellular hydrostatic pressure. This reduction can lead to decreased cellular water potential but, cellular activity will only be slightly altered by this pressure reduction and changing this by 20 atmospheres will only change the water activity by 3 or 4 %. Burton (1989) described the vapour pressure gradient between the surface of fruits or vegetables and the air in which they are in contact. A major proportion of fresh fruits, vegetables and flowers is water which will pass from cell to cell and eventually to the surface where it evaporates using latent heat from the commodity. This principle is used in vacuum cooling of fruits and vegetables. Some fruits and vegetables are coated with a waxy cuticle or, in the case of some root crops, a periderm that restricts this water loss. The permeability of fruits and some vegetable tissue to gas exchange is also affected by intercellular spaces, cuticle composition as well as the presence of stomata, lenticels and hydrothodes. Ben-Yehoshua and Rodov (2003) and Laurin et al. (2006) commented that it is likely that desiccation of fruit and vegetable under hypobaric storage is related to an increase in transpiration rate enhanced by the properties and action of stomata, lenticels, cuticle and epidermal cells. In grapes the epidermis does not contain a significant number of functional stomata, therefore water loss occurs mostly through the cuticle, which in turn restricts water loss. Stomata have been shown to be affected by storage conditions. For example, after 96 h storage cucumbers under hypobaric conditions had significantly more open stomata than those under atmospheric pressure (Laurin et al. 2006). They also commented that it is likely that desiccation under hypobaric conditions is due to an increase in transpiration rate enhanced by the properties and action of stomata as well as moisture being more volatile at reduced atmospheric pressure.

Diseases

Although the high humidity maintained in hypobaric stores is generally suitable for fungal growth and decay development, there are several reports that show hypobaric conditions can reduce decay. Many authors have pointed out that hypobaric conditions can retard or limit pathogen growth (Burg and Kosson 1983; Goszczynska and Ryszard 1988; Lougheed et al. 1978; Chau and Alvarez 1983). Inhibition of the germination and growth of fungal spore at low O_2 levels was demonstrated in the early studies of Brown (1922). Couey et al. (1966) showed that postharvest decay in strawberry fruit was reduced at O_2 levels of 0.5 % or less and they demonstrated a direct effect of such low O_2 environments on growth of mycelium and sporulation. Less pathogenic breakdown was observed in cranberries under hypobaric conditions of 76 mm Hg than in fruit stored in atmospheric pressure. Similarly, Chau and Alvarez (1983) reported that papaya inoculated with *C. gloeosporioides* developed less infection when stored under about 15 mm Hg than when stored under normal atmospheric pressure. Apelbaum and Barkai-Golan (1977) showed that the degree of inhibition of fungal growth in hypobaric stores increased with the reduction in pressure below 150 mm Hg.

They also reported that hypobaric pressure had direct fungistatic effects on spore germination and mycelium growth of various storage fungi. In a review, the control of postharvest diseases by maintaining O_2 partial pressure in the region of $0.1–0.25 \pm 0.008$ % was reported by Burg (2004). However, these low O_2 levels could damage the fruit. In contrast, studies by Bangerth (1974) with various fruit and vegetables, including tomatoes, peppers and cucumbers, found a high incidence of decay after storage at hypobaric pressure. In order to suppress decay, he recommended a combination of hypobaric storage with postharvest fungicidal treatments. Barkai-Golan (1977) reported that in in vitro studies, different fungi responded differently to reduced pressures. For example storage under 100 mm Hg inhibited spore germination of *P. digitatum* compared to 760 mm Hg while inhibition of spore germination of *B. cinerea* and *Alternaria alternata* occurred only at 50 mm Hg. However, 50 mm Hg had no effect on the germination of *Geotrichum candidum* spores, but reducing the pressure to 25 mm Hg totally prevented spore germination of *P. digitatum, B. cinerea* and *A. alternata* but had almost no effect on the germination of *G. candidum*. Transfer of inhibited cultures from hypobaric to atmospheric pressure resulted in renewed growth, suggesting that there was no irreversible damage to the fungi (Alvarez 1980, Alvarez and Nishijima 1987). In contrast, the effectiveness of short hypobaric treatments against postharvest diseases was investigated by Romanazzi et al. (2001) who found that it reduced fungal infections in sweet cherries, strawberries and table grapes. Adams et al. (1976) investigated the effects of a range of pressures (760–122 mm Hg) on the growth of *P. expansum* and *P. patulum* and their production of the toxin patulin. They demonstrated that the amount of sporulation decreased with reduction in pressure, but mycelial growth was similar for 456 and 357 mm Hg and patulin production was lower at the lower pressures.

Insects

Disinfestation of fruits and vegetables by exposure to hypobaric conditions during export has been described by Burg (2010), Chen et al. (2005), Davenport et al. (2006), Johnson and Zettler (2009), Mbata and Philips (2001) and Navarro et al. (2001, 2007). Effective control of insects has been observed at O_2 concentrations of less than 6.6 %, and especially at 0.15–0.30 % (Burg 2004). Burg (2004) and Aharoni et al. (1986) gave the optimal condition for transporting many tropical fruits at 13 °C and where the pressure in the container could be reduced to 15–20 mm Hg this would kill most insects infesting the fruit. They claimed that in these conditions 98 % of fruit fly eggs and larvae were killed within one week and all of them by the 11th day and all the green peach aphids on wrapped head lettuce in 2½ days at 2 °C. Insect mortality under hypobaric storage is predominantly caused by low O_2 concentrations (Navarro and Calderon 1979), although the low humidity that could be generated in a hypobaric system could also enhance its lethal effect on insects (Jiao et al. 2012a, Navarro 1978). When insects were placed into a hypoxic environment for a sufficient duration, adenosine triphosphate production was reduced, resulting in increasing membrane phospholipid hydrolysis (Herreid 1980). Cell and mitochondrial membranes then become permeable, causing cell damage or death (Mitcham et al. 2006). A patent was taken out in 2005 by Timothy K Essert and Manuel C Lagunas-Solar of the University of California in the USA (PCT/US2004/013225) on "A method and system for disinfecting and disinfecting a commodity, such as a perishable agricultural commodity, by treatment with an environment of low oxygen/high ballast gas with cycled pressure changes that overwhelm and damage the respiratory system of the insect without damaging the host commodity".

Contamination

Hypobaric conditions have been showed to be effective in enabling different crops to be stored together without mutual contamination. For example, carrots exposed to ethylene can synthesise isocoumarin, which gives them a bitter taste (Lafuente et al. 1989). Keeping apples, cabbages and carrots together in a store at 2 °C and 60 mm Hg resulted in no isocoumarin detected in the carrots despite the assumed presence of ethylene from the apples (McKeown and Lougheed 1981). A sensory panel compared those carrots with those stored in O_2 levels below 2 % and bags of slaked lime (calcium hydroxide) to absorb CO_2 and found that those from the hypobaric storage were superior. Burg (2004) described an experiment where bananas and apples were stored at 14.4 °C, either together or separately and either under 760 or 80 mm Hg. The bananas stored with tomatoes under 760 mm Hg initiated to ripen because of the ethylene given out be the tomatoes while the bananas under 80 mm Hg did not initiate to ripen.

Horticultural Commodities

Apples

Optimum hypobaric conditions appear to vary with cultivars. In Korea, Kim et al. (1969, quoted by Ryall and Pentzer 1974) reported that optimum conditions for Summer Pearmain were 200 mm Hg and for Jonathan it was 100 mm Hg. Bubb and Langridge (1974) found no extension in the storage life of the apple cultivar Cox's Orange Pippin at 3.3 °C under 380 mm Hg. However, they found that storage of Tydeman's Late Orange at 3.3 °C had reduced respiration rates and ethylene production under 76 mm Hg (0.1 atmosphere) compared to those stored under atmospheric pressure or under 380 mm Hg. Bubb (1975b) harvested Cox's Orange Pippin apples in September and compared the following storage conditions in 3.3 °C: air at atmospheric pressure, controlled atmosphere conditions of 2 %

$O_2 + 0$ % CO_2 at atmospheric pressure and hypobaric conditions of 35–40 and 70–80 mm Hg. They found that the onset of ethylene biosynthesis was delayed by 50 days in the controlled atmosphere stored fruit, by 70 days in the hypobaric stored fruit at 70–80 mm Hg and by 100 days under the 35–40 mm Hg all compared to those stored in air in atmospheric pressure. Respiration rate was about half the level in the fruit that had been stored under hypobaric conditions and slightly more in those that had been stored under controlled atmosphere compared to those that had been stored in air under atmospheric pressure. This effect continued when fruit were removed to 10 °C in air for 3 weeks. Those under hypobaric storage had higher weight losses than the other fruit but those that had been stored under 35–40 mm Hg were firmer that those in air or controlled atmosphere storage. In a subsequent experiment, weight loss was reduced to 0.3 % per month by improving the humidification system. The fruit from both hypobaric conditions were assessed as having poorer flavour and a "denser texture" than those that had been stored in air or controlled atmosphere conditions. Bubb (1975c) also compared the storage of Cox's Orange Pippin apples at 3.3 °C in air with controlled atmosphere storage under 2 % $O_2 + 0$ % CO_2 and 1 % $O_2 + 0$ % CO_2 at atmospheric pressure and under hypobaric storage in 25–30 and 50–60 mm Hg. He found that respiration rate of those under hypobaric storage was some 50 % higher than those stored under controlled atmosphere storage, but ethylene production was lower in those stored under 25–30 mm Hg compared to those under controlled atmosphere storage in 1 % $O_2 + 0$ % CO_2 but ethylene production was inhibited in fruit under 50–60 mm Hg only until mid December after which the production rate rose sharply. The ethylene production rate of the fruit under 2 % $O_2 + 0$ % CO_2 were similar to the latter but the rise in ethylene production rate was much slower. Little physiological damaged was observed on any fruit until the end of April when all samples showed lenticel rotting, with fruit under 50–60 mm Hg being the worst, and slight core flush and breakdown with 25–30 mm Hg being the worst for both disorders. Sound fruit from all treatments were transferred to 12 °C at that time and those that had been stored under 25–30 mm Hg continued to show depression in ethylene production for 2–3 weeks. Sharples and Langridge (1973) found that during storage of Cox's Orange Pippin at 3.3 °C with airflow of 15 L h^{-1} had more lenticel blotch pit for those stored under 380 mm Hg compared to those at atmospheric pressure. However they reported that there was only 20 % breakdown in the fruit stored under 380 mm Hg compared to 50 % in those under atmospheric pressure in apples from poor keeping quality orchards and they commented that this effect may have been related to higher weight loss from the fruit under hypobaric conditions. Bangerth (1984) reported that apples stored under hypobaric pressure of 51 mm Hg never produced autocatalytic ethylene or developed a respiratory climacteric during 11 months storage. Only a slight decrease in fruit firmness was measured during that time. When ethylene was continuously supplied into the hypobaric containers, a considerable response was observed at the beginning of the storage period, but later the effect of ethylene was only marginal. He also found that there was no diminished response to ethylene in storage under 51 mm Hg, whatever the storage temperature tested. There was a similar decrease in sensitivity to ethylene in terms of respiration rate, softening and volatile flavour substances after shelf life evaluation with the fruits that had been stored under hypobaric pressures for 2.5, 5, 7 and 10 months.

Storing apples under hypobaric conditions resulted in delayed softening, control of physiological disorders and reduced decay development, which resulted in extended shelf life after removal from storage (Laugheed et al. 1978; Dilley et al. 1982). Wang and Dilley (2000) also found that hypobaric conditions could prevent the development of scald. They reported that apples of the cultivars of Law Rome and Granny Smith that were placed under hypobaric conditions of 38 mm Hg within 1 month of harvest did not develop scald during storage at 1 °C but if there was a delay in establishing hypobaric conditions of 3 months then scald developed as it did on fruit that were stored under atmospheric pressure throughout.

Jiao et al. (2013) suggested that exposure to 10 mm Hg at 10 °C and greater than 98 % r.h. had potential as an alternative disinfestation treatment against codling moth in apples and 15 days exposure to the cultivar Red Delicious had no detrimental effect on fruit quality. They studied eggs, 2nd to 3rd instar larvae, 5th instar larvae and pupae and found that the 5th instar larvae were the most tolerant stage for codling moth exposed to the treatment.

Asparagus

At 0 °C and 20, 40 or 80 mm Hg Dilley (1990) found that spears could be kept in marketable condition for 4–6 weeks with better ascorbic acid retention at 20 mm Hg that at the other hypobaric conditions. Li and Zhang (2006) reported an extension in postharvest life of green asparagus in storage under 112.5 ± 37.5 mm Hg. Wenxiang et al. (2006) found that at room temperature storage life of asparagus was 6 days, in refrigerated storage it was 25 days and in refrigerated storage with hypobaric conditions it was up to 50 days. They also reported that the spears under hypobaric storage had a lower respiration rate, lower losses of chlorophyll, ascorbic acid, titratable acidity, total soluble solids, reduced malondialdehyde accumulation and improved sensory quality. McKeown and Lougheed (1981) found that in storage at 3 °C and near saturation humidity under 61 mm Hg and also under 2 % O_2 + 0 % CO_2 controlled atmosphere asparagus spears remained firm and green for 42 days while those in air were senescent. However, spears from all three treatments showed a disorder which resembled chilling injury. They concluded that there appears to be limited potential for the storage of asparagus at 3 °C. The storage life fresh green asparagus under hypobaric pressure of 266–342 mm Hg was extended to 50 days compared to 25 days refrigerated storage and only 6 days under room temperature both at atmospheric pressure. Furthermore, their respiration rate was lower and it prevented loss of chlorophyll, vitamin C and acidity, improved sensory qualities and delayed postharvest senescence when stored under hypobaric conditions (Li et al. 2006).

Li (2006) stored the cultivar UC800 in a hypobaric chamber in a vacuum pressure of about 35.40 kPa, storage temperature of -2 °C and humidity 85 % $\pm$ 5 % r.h. The three-stage hypobaric storage technology of green asparagus the vacuum pressure were respectively -10.0 kPa in the first stage, 20.0 kPa in the second stage and 35.0 kPa in the third stage (sic). The result showed that the effect of three-stage hypobaric storage was obviously better than the normal hypobaric storage. Atmosphere cold storage used as the control, The result indicated that the three-stage hypobaric storage condition could significantly ($p < 0.05$) inhibit the degradation of sugar, soluble protein, ascorbic acid and total acid, decrease the senescence index and improve the commodity rate of green asparagus compared with the atmosphere cold storage. The three-stage hypobaric storage condition could significantly ($p < 0.05$) inhibit respiratory intensity and ethylene emission, increase the activities of superoxide dismutase and catalase, decrease the accumulation of superoxide anion and hydrogen peroxide, reduce the damage of cell membrane.

Avocado

Burg (2004) summarised his work over many years on the effects of hypobaric storage on avocados. The cultivar Choquette stored at 14.4 °C under atmospheric pressure started to ripen in 8–9 days and they were fully ripe in 14 days. Softening of those under 40–101 mm Hg began softening after 25 days and when they were transferred to 20 °C under atmospheric pressure all fruit developed normal taste with no internal blackening or decay. He subsequently found that in storage at 12.8 °C hypobaric conditions of 101–152 mm Hg was better than at 40–81 mm Hg and in later work he reported that 21 mm Hg was optimal at 10 °C. With the cultivar Waldin, he reported that in storage at 10 °C their postharvest life was improved as the pressure was lowered from 101 to 152 mm Hg down to 60–81 mm Hg with the fruit remaining firm for 30 days at 61–81 mm Hg compared to 12–16 days at atmospheric pressure. He reported similar results for avocados in storage at 12 °C but all fruit ripened quicker.

Spalding and Reeder (1976) compared storage of Waldin at 7.2 °C and 98–100 % r.h. for 25 days at atmospheric pressure in air with controlled atmosphere storage under 2 % O_2 and 10 % CO_2 or 2 % O_2 and 0 % CO_2 and two hypobaric storage conditions in 91 mm Hg, one with added CO_2 at 10 %. After storage all the fruit was ripened at 21.1 °C. They found that 92 % of the fruit stored in the controlled atmosphere of 2 % O_2 and 10 % CO_2 were acceptable and all those in the hypobaric conditions of 91 mm Hg plus 10 % CO_2 while none of the fruit in the other treatments were acceptable. The factors that affected acceptability were anthracnose disease (*C. gloeosporioides*) and chilling injury, both of which were completely absent in fruit stored under 91 mm Hg plus 10 % CO_2. They defined acceptable fruit as having good appearance, free of moderate or severe decay and chilling injury and had no off-flavours. They also found no stem-end rot (*Diplodia natalensis*) directly after storage, but after ripening at 21.1 °C again no stem-end rot was detected except low levels on those that had been stored under 91 mm Hg and higher levels in those that had been stored under 2 % O_2 and 10 % CO_2. "Black pitted areas developed in lenticels during softening of avocados stored at atmospheric pressure or hypobaric plus 10 % CO_2. However, pitting was slight and was not considered to be objectionable to the average consumer. Tissue from the infected areas contained *Pestalotia* spp. fungus". From this they concluded that high CO_2 was necessary for the successful storage of avocados since the hypobaric system would have reduced the partial pressure of O_2 91 mm Hg would be about 2.5 %. This conclusion for high CO_2 to be necessary for optimising storage of avocados under controlled atmosphere storage is borne out by many other workers. However, Burdon et al. (2008) found that the inclusion of CO_2 at 5 % under controlled atmosphere storage retarded fruit ripening but stimulated rot expression and they concluded that it should not be used for controlled atmosphere storage of New Zealand grown Hass. With the avocado cultivar Lula, Burg (1969) reported that in storage at 7.2 °C they began to soften within 21 days and all were soft after 41 days under atmospheric pressure while under 81–122 mm Hg it took 88 days for the fruit to begin to soften. In another experiment at 8 °C and 40–81 mm Hg the fruit remained firm for 75–100 days compared to 23–30 days under atmospheric pressure. When they were stored at 61 mm Hg for 102 days they became eating ripe within 3–4 days when they were transferred to 26.7 °C at atmospheric pressure (Burg 2004). Spalding and Reeder (1976) compared storage of Lula at 10 °C and 98–100 % r.h. for 6 weeks at atmospheric pressure in air, controlled atmosphere storage under 2 % O_2 and 10 % CO_2 and hypobaric storage under 76 and 152 mm Hg. After storage all the fruit was ripened at 21.1 °C and they found that 70 % of the fruit stored under the controlled atmosphere were acceptable and none of the fruit in the other treatments were acceptable, which was mainly due to chilling injury symptoms and decay due to anthracnose. Lula stored under 2 % O_2 with 10 % CO_2 under atmospheric pressure were acceptable after softening and also this controlled atmosphere mixture inhibited the development of decay and chilling injury confirming the finding that CO_2 is essential under controlled atmosphere storage (Spalding and Reeder 1972, 1975).

With the cultivar Booth 8, Burg (2004) reported that after storage at 4.4 °C under 40, 61, 81 or 122 mm Hg for 30 days they ripened in 2–3 days when transferred to 20 °C under atmospheric pressure but had a poor flavour. In another experiment waxed Booth 8 were stored at 7.8–10 °C where they began to ripen in 8–22 days under atmospheric pressure, while under 40–81 mm Hg they did not soften during 50 days storage but ripened without skin darkening when they were transferred to 23.9 °C under atmospheric pressure. After 64 days storage at 7.8–10 °C under 40–81 mm Hg fruit were still firm but they did not ripen to an acceptable quality when transferred to 23.9 °C under atmospheric pressure.

With the cultivar Hass it was reported that in storage at 5 °C under atmospheric pressure fruits softened within 30 days while under 15–40 mm Hg they were still hard and almost half of those under 61 mm Hg had begun to ripen (Cicale and Jamieson 1978 quoted by Burg 2004). In storage under 40 mm Hg fruit began to soften after 38–45 days but under 15–21 kPa they remained firm and when removed to ambient conditions they ripened in 5.4 days, which was the same period as freshly harvested fruit. In another trial at 6 °C for 35 days, fruit under atmospheric pressure lost 1.2 % in weight at pressures ranging from 61 to 203 mm Hg compared to 5.7 % under atmospheric pressure. After 70 days storage the weight losses were 1.7 % for 81 and 101 mm Hg and 3 % for 61 mm Hg. They commented that their best results were storage at 61 mm Hg since this retarded softening and fruits ripened normally when transferred to 14 °C under atmospheric pressure and lower storage pressures resulted in higher desiccation.

Spores of *Glomerella cingulata* (which causes anthracnose) germinate on the surface of avocado fruit in the field and form appressoria. The fungus then remains quiescent until antifungal dienes in the skin of fruit breakdown due to degradation by lipoxygenase activity. Breakdown of the dienes has been shown to be delayed by various treatments including hypobaric storage (Prusky et al. 1983, 1995). Previous studies on hypobaric storage of avocados suggested that atmospheres both low in O_2 and high in CO_2 are required for successful suppression of anthracnose development (Spalding and Reeder 1976).

Apricots

Salunkhe and Wu (1973) and Haard and Salunkhe (1975) found that storage life of apricots could be extended from 53 days in cold storage to 90 days in cold storage combined with reduced pressure of 102 mm Hg. They found that hypobaric storage delayed carotenoid production, but after storage carotenoid, sugar and acid levels were the same as those that had been in cold storage at atmospheric pressure.

Bamboo Shoots

Huiyun Chen et al. (2013a) stored freshly harvested bamboo shoots (*Phyllostachys violascens*) at 2 ± 1 °C under various hypobaric conditions (101, 75, 50 and 25 kPa) for 35 days. They found that under 50 kPa there were reduced accumulations of lignin and cellulose in their cells. They also found that it inhibited ethylene production, reduced the rate of accumulation of malondialdehyde (MDA) and hydrogen peroxide, and maintained significantly higher activities of superoxide dismutase (SOD), catalase (CAT) and ascorbate peroxidase (APX), but restrained the activities of phenylalanine ammonia-lyase (PAL) and peroxidase (POD). They therefore concluded that the delay in flesh lignification was due to maintenance of higher antioxidant enzymes activities and reduced ethylene production.

Bananas

In his review Burg (2004) reported that at 13.3–14.4 °C the banana varieties Valery and Gros Michel ripened in 10 days under atmospheric pressure but remained green for 40–50 days at 152 mm Hg, but mould could develop. He also reported that Valery at 13.3 °C remained green for more than 105 days in storage under 6.4, 9.5 or 12.7 kPa and ripened normally when they were transferred to atmospheric pressure and exposed to exogenous ethylene. There were no deleterious effects on flavour or aroma. Burg (1969) stored bananas under 49, 61, 76, 101, 122 and 167 mm Hg for 30 days and found they lost 1.1–3.6 % in weight with the higher losses at the higher pressures because of their higher respiration rates. All the fruit were still green and there were no apparent differences between the different hypobaric conditions. Bangerth (1984) found that there was no diminished response to ethylene in storage under 51 mm Hg, whatever the storage temperature tested. Apelbaum et al. (1977a) stored the variety Dwarf Cavendish at14 °C at 81 and 253 mm Hg and atmospheric pressure and found that they began to turn yellow after 30 days under atmospheric pressure, 60 days under 253 mm Hg and were still dark green under 81 mm Hg. When they were subsequently transferred to atmospheric pressure at 20 °C and exposed to exogenous ethylene they all ripened to a good flavour, texture and aroma. Bangerth (1984) successfully stored Cavendish at 14 °C under 51 mm Hg for 12 weeks and found that when they were subsequently ripened in 50 µl L-1 exogenous ethylene they were the same quality as freshly harvested fruit or those that had been stored under atmospheric pressure.

Hypobaric storage has also been shown to delay the speed of ripening of bananas that have been initiated to ripen. Quazi and Freebairn (1970) found that fruit that had been initiated to ripen by exposure to exogenous ethylene for 2–5 h longer than required to initiate ripening, did not ripen during hypobaric storage but began to ripen within 1–2 days after transfer to ambient atmospheric pressure and had good eating quality and texture. However fruit that had been initiated to ripen by exposure to ethylene with exposure time of 16 h longer than required did ripen during hypobaric storage again with good eating quality and texture. Liu (1976) initiated Dwarf Cavendish to ripen by exposure to 10 µl L-1 ethylene at 21 °C. They then stored them at 14 °C for 28 days under hypobaric storage of 51, 79 mm Hg or controlled atmosphere storage of 1 % O_2 + 99 % N2. All fruit remained green and firm and continued to ripen normally after they had been removed to ripening temperature in atmospheric pressure. Quazi and Freebairn (1970) showed that high CO_2 and low O_2 delayed the increased production of ethylene associated with the initiation of ripening in bananas, but the application of exogenous ethylene was shown to reverse this effect. Wade (1974) showed that bananas could be ripened in atmospheres of reduced O_2, even as low as 1 %, but the peel failed to degreen, which resulted in ripe fruit which were still green. Similar effects were shown at O_2 levels as high as 15 %. Since the degreening process in Cavendish bananas is entirely due to chlorophyll degradation (Seymour et al. 1987; Blackbourn et al. 1990), the controlled atmosphere storage treatment was presumably due to suppression of this process. Hesselman and Freebairn (1969) showed that ripening of bananas, which had already been initiated to ripen by ethylene, was slowed in low O_2 atmospheres.

Beans

Green bean pods (*Phaseolus vulgaris*) could be kept in good condition in storage for 30 days at reduced pressure compared to 10–13 days in cold storage alone (Haard and Salunkhe 1975). Spalding (1980) reported that the cultivars McCaslan 42 pole beans and Sprite bush beans stored better at 7 °C under 76 and 152 mm Hg for 2 weeks than similar beans stored at 760 mm Hg. Burg (1975) found that beans stored at 7.2 °C and 60 mm Hg were in excellent condition after 26 days compared to those stored at the same temperature at 760 mm Hg that were shrivelled and in poor condition. Knee and Aggarwal (2000) found water soaked lesions appeared on green beans kept in vacuum containers at 380 mm Hg, however, this effect is commonly associated with water condensation (Thompson 2015).

Hypobaric storage has been shown to be effective on dried beans. Berrios et al. (1999) reported that the combined effect of refrigeration and hypobaric storage demonstrated potential for maintaining the fresh quality of black beans (*P. vulgaris*) in storage for up to 2 years. Black beans stored at 4.5 °C and 50–60 % r.h. and hypobaric pressure of 125 mm Hg exhibited quality factors characteristic of fresh beans, such as shorter cooking time, smaller quantities of solids loss, lower leaching of electrolytes and lower percentage of hard-shell than beans stored at 23–25 °C and 30–50 % r.h. At 4.5 °C beans stored under 125 mm Hg had a germination rate of 93 % while those stored at atmospheric pressure had 72 %. Beans stored in ambient conditions exhibited hard-to-cook.

Beets

In general, the weight loss after storage at 1 °C under 61 mm Hg and near-saturation humidity and the controlled atmosphere storage under 2 % O_2 + 0 % CO_2 was less than that which occurred after air storage. The beets also retained a fresh appearance after holding under 61 mm Hg and sensory evaluation indicated that those held under 61 mm Hg were similar to those held in air, while beets held under 2 % O_2 + 0 % CO_2 had a lower rating and off-flavours (McKeown and Lougheed 1981).

Blueberries

Al-Qurashi et al. (2005) found that the cultivar Rabbiteye stored at 1.0 atmospheric pressure (13.5 mm Hg sic) lost less weight, were firm, developed less decay and did not show any shrivelling during storage at 4 °C for 28 days compared to those stored at atmospheric pressure (*sic*) as a control. Borecka and Pliszka (1985) observed that blueberries stored under 38 mm Hg tasted good, contained less acid and had lower total soluble solid than other treatments. Burg (2004) reported that in storage at 0–2 °C those under atmospheric pressure spoiled within 4 weeks while those under hypobaric storage ranging from 81 to 203 mm Hg their storage was limited to 6 weeks. This was reported to be because of mould growth. Burg (2004 also quoted David Dilley 1989) who reported similar results on reduction in mould on the cultivar Jersey during 44 days storage, with 87 % decay for those under atmospheric pressure and only some 11 % for those under 21 mm Hg.

Broccoli

Broccoli heads were kept for 4 days at 1.1 °C then at 0 °C and 10 mm Hg or atmospheric pressure for 21 days. They were then assessed for quality after a shelf life at 10 °C. The ones that had been stored under atmospheric pressure had 60–90 % yellowing and those that been in hypobaric storage had 40 % yellowing with no off odours or off-flavours (Burg 2004).

Brussels Sprouts

Burg (2004) recommended that storage of Brussels sprouts under 10 mm Hg should be tested since it could prevent the growth of microorganism that could result in postharvest diseases, although Ward (1975) found no improvement in storage under 76 mm Hg and high humidity.

Cabbages

Cabbages, like other leaf vegetables, become chlorotic when exposed to ethylene. It was reported that white cabbages were successfully stored with apples at 0 °C and 60 mm Hg (McKeown and Lougheed 1981) presumably by limiting the effect of ethylene produced by the apples or removing the ethylene from the container before it could have an effect. Onoda et al. (1989) found that cabbages retained better appearance and had lower weight loss when stored in cycles between 100 and 300 mm Hg (with no humidification) than those stored at a constant atmospheric pressure.

Capsicum

Kopec (1980) compared storage of capsicum at atmospheric pressure with storage under 77 mm Hg and found that hypobaric storage retarded ripening but once ripening started it progressed at the same rate as in atmospheric pressure. However, capsicums are classified as non-climacteric (Bosland and Votava 2000) so it is difficult to interpret these results since climacteric fruits do not ripen. Perhaps what is being referred to is change of colour from green. Hughes et al. (1981) experimented with hypobaric stored capsicums at 8.6–9.0 °C and 82–90 % r.h. under 38, 76 or 150 mm Hg. They found that they had a significantly higher weight loss during storage than those stored at normal pressure. This was undoubted due to the comparative low humidity, which should have been as close as possible to saturation. However, all had similar levels of "sound" fruits after 20 days within the range of 60–72 %. Hypobaric stored capsicums did not have an increased subsequent storage life compared to those stored under atmospheric pressure at the same temperature (Table 3.3). However, Burg (1975) found that capsicums could be kept for seven weeks at 7.2 °C under hypobaric storage without loss of colour or "crispness". Burg (2004) reported that under storage at 7.2 °C and 80 mm Hg capsicums were in excellent condition after 28 days while those stored under atmospheric pressure began to deteriorate after 16 days and were in poor condition after 21 days. Even after 46 days capsicums under hypobaric storage were considered to be marketable except for a trace of mould on the cut stem. It was concluded that decay was the limiting factor in hypobaric storage. Jamieson (1980 quoted by Burg 2004) found that after 50 days 87 %, 80 %, 47 % and 0 % were saleable (sic) from 80, 40, 15 mm Hg and atmospheric pressure respectively in experiments in the Grumman Allied Industries laboratory. Bangerth (1973) stored the cultivar Neusiedler Ideal at 10–12 °C for 23 days and found that those under 75 mm Hg were firmer, greener and had slightly higher ascorbic acid content and lower ethylene production than those stored under atmospheric pressure.

Table 3.3 The effects of hypobaric conditions during storage and after removal on the mean percentage weight loss of fruit as a percentage of their original weight and on the mean percentage of sound marketable fruit. *Source* modified from Hughes et al. (1981)

Atmospheric pressure (mm Hg)	20 days hypobaric storage at 8.8 °C		7 days subsequent shelf life in 760 mm Hg pressure at 20 °C	
	Weight loss %	Marketable fruit %	Weight loss %	Marketable fruit %
760	0.03	68	0.87	43
152	0.16	72	0.76	46
76	0.22	68	1.00	36
38	0.15	60	1.10	42

Cauliflowers

Cauliflowers were kept for 4 days at 1.1 °C then 0 °C and 10, 20 or 40 mm Hg or atmospheric pressure for 21 days. They were then assessed for quality after a shelf life at 10 °C by Burg (2004). At the end of the shelf life test the ones that had been stored under atmospheric pressure had leaves that were yellow and dry and easily abscised with minimal handling. Those that been in hypobaric storage had some yellowing but remained firmly attached when handled and had a superior appearance especially those that had been stored at 10 or 20 mm Hg. However, Ward (1975) found no improvement in storage under 76 mm Hg and high humidity.

Cherries

The cultivar Bing stored well for 93 days in experimental hypobaric chambers at 102 mm Hg at 0 °C. This was up to 33 days longer than those stored at 0 °C under atmospheric pressure. However, their pedicels stayed green for only 60 days and pedicel browning can affect their marketability. Hypobaric conditions delayed chlorophyll and starch breakdown in fruit as well as carotenoid formation and the decrease in sugars and total acidity (Salunkhe and Wu 1973). Patterson and Melsted (1977) reported that cherries could be stored for 6–10 weeks (depending on their condition at harvest) at 41–203 mm Hg with good retention of colour and brightness and delayed disease development. They also found that they stored equally well under controlled atmosphere storage with high CO_2 but under hypobaric conditions there was some problems with desiccation especially at 41 mm Hg. Controlled atmosphere storage recommendations include: −1 to −0.6 °C with 20–25 % CO_2 and 0.5–2 % O_2 helped to retain fruit firmness, green pedicels and bright fruit colour (Hardenburg et al. 1990), −1.1 with 20–25 % CO_2 with 10–20 % O_2 (SeaLand (1991), 20–30 % CO_2 reduced decay (Haard and Salunkhe 1975) and 0–5 °C with 10–12 % CO_2 and 3–10 % O_2 (Kader 1989).

The effectiveness of short hypobaric treatments against postharvest rots was investigated by Romanazzi et al. (2001) who found that the cultivar Ferrovia exposed to 0.50 atmosphere for 4 h had the lowest incidence of *B. cinerea*, *Monilinia laxa* and total rots. Romanazzi et al. (2003) also found that the combination of spraying with chitosan at 0.1, 0.5 or 1.0 % 7 days before harvest and storage under 0.50 atmosphere for 4 h directly after harvest effectively controlled fungal decay of sweet cherries during 14 days storage at 0 ± 1 °C, followed by a 7 day shelf life. Fungi associated with rots included brown rot (*M. laxa*), grey mould (*B. cinerea*), blue mould (*P. expansum*), Alternaria rot, (*Alternaria*, sp.) and Rhizopus rot (*Rhizopus* sp.).

Cranberries

Storing cranberries (*Vaccinium macrocarpon*) under 80 mm Hg resulted in a lower respiration rate and ethylene production as well as an extended shelf life compared to those stored under atmospheric pressure (Pelter 1975 quoted by Al-Qurashi et al. 2005). Lougheed et al. (1978) explained that ethylene and respiration rate of cranberries stored under 80 mm Hg (0.1 atmosphere) decreased compared to fruit stored under atmospheric pressure.

Cucumbers

Hypobaric storage of cucumbers at 0.1 atmosphere (76 mm Hg) extended the storage life to 7 weeks, compared with 3–4 weeks in cold storage (Bangers 1974) and reduced respiration rate by 67–75 %. Burg (2004) reported that storage at 7.2 °C resulted in chilling injury and reducing the pressure to 100, 120 or 160 mm Hg reduced chilling injury. However, under 80 mm Hg there were no symptoms after 7 weeks but chilling injury occurred within 1 or 2 days when they were removed to ambient conditions. Cucumbers were placed at 71 mm Hg for 6 h in the dark at 20 °C to simulate air flight transportation and placed subsequently in cold storage facilities at 101 mm Hg at 20 °C and 70 % r.h. for 7 days. Results showed that cucumbers exposed to the hypobaric conditions had significantly more open stomata, compared to those at atmospheric pressure, after 96 h of subsequent storage (Laurin et al. 2006).

Currants

Bangerth (1973) stored blackcurrants, redcurrants and whitecurrants (all *Ribes sativum*) at about 3 °C either in atmospheric pressure or 76 mm Hg for up to 38 days. Those under the hypobaric conditions retained their ascorbic acid content better than those under atmospheric pressure and also had higher sugar content, less decay and a better taste. Percentage spoiled berries after 38 days storage were 21–37 % for those under atmospheric pressure and only 0.5–5 % for those under hypobaric conditions.

Cut Flowers

Hypobaric storage has been extensively tested on the postharvest life many species of cut flowers. For example, Staby et al. (1984) reported that at 2.9 ± 1.1 °C and over 90 % r.h. roses could be stored for up to 2 weeks at atmospheric pressure but

Table 3.4 Vase life of roses stored in water at 4 °C for different periods at different atmospheric pressures where vase life was 7 days for freshly harvested flowers. *Source* modified from Bangerth (1973)

Storage time (weeks)	Vase life (days)		
	Atmospheric pressure	75 mm Hg	40 mm Hg
2	4.0	7.5	–
4	0.5	6.1	5.1
6	0	6.5	1.1

up to 4 weeks at 10–35 mm Hg. After storage the roses still retained at least 61 % of the vase life of flowers that had been freshly harvested. However, they reported that leaf disorders developed on the hypobaric stored flowers. Burg (2004) stored roses of the cultivar Sweetheart at 1.7 °C with their stems cut in water and found that under atmospheric pressure their vase life was 14–18 days while under 40 mm Hg the buds were still tight and were in excellent condition. However, under 60 and 80 mm Hg leaves were slightly wilted and the disorder 'bent neck' occurred in the flowers. Bangerth (1973) also showed that roses of the cultivar Baccara stored under hypobaric conditions had longer shelf life (Table 3.4).

Dilley and Carpenter (1975) stored carnations at 0 °C in humidified air at 50 mm Hg for 9 weeks and found a marked extension in their longevity. Failure of carnations harvested at the bud stage to open properly was completely prevented under hypobaric storage and the vase life of buds or fully open carnations in holding solutions was generally as good as or better than freshly harvested flowers. In Denmark, Brednose (1980) reported that roses could be kept longer under hypobaric conditions than under atmospheric conditions. The cultivars Belinda and Tanbeede were harvested at the bud stage with two sepals open and wrapped in polyethylene and stored for one month at 2 °C and 98 % r.h. in either 24 mm Hg or atmospheric pressure with an air exchange of one volume per hour. On removal the flowers from the hypobaric store kept fresh for 7 days. The experiment was repeated with other cultivars but did not produce such good results and leaf injuries, spots and wilting of the leaves was observed. Staby (1976) reported that "bud cut" chrysanthemums could be stored for 6 weeks at 3 °C under 25 mm Hg and had double the vase life of those stored under atmospheric pressure. Burg (2004) also reported that "bud cut" chrysanthemums stored for 6 weeks at 0–1.7 °C in polyethylene lined boxes were in excellent condition when stored under 10–25 mm Hg with no loss in their ability to open and little or no loss of subsequent vase life. He also reported that waxed paper was placed between the blooms to prevent water spotting due to condensation, which had been reported to occur on carnations and roses. He also gave a general evaluation of storage of some cut flowers. For carnations, *Protea* spp. and roses there was no benefit of controlled atmosphere storage and their postharvest life in cold storage at atmospheric pressure was 21–42 days, less than 7 days and 7–14 days respectively. But under hypobaric conditions these times were increased to 140 days, over 30 days and 42–56 days respectively. In his review (Burg 1975), he gave the maximum storage time under hypobaric conditions as 91 days for carnations, 56 days for roses, 28–35 days for red ginger, 41 days for the vanda orchid Miss Joaquim, 21–28 days for chrysanthemums, 42–56 days for antirrhinums and 30 days for gladiolus.

Grapes

Burg (1975) found that hypobaric storage extended the postharvest life of grapes from 14 days in cold storage in air to as much as 60–90 days. Burg (2004) subsequently reported that the cultivar Red Emperor could be stored for 90 days under hypobaric pressures of 30, 65, 80 or 160 mm Hg at 1.7–4.4 °C compared to 21–56 days at −0.5–0 °C at atmospheric pressure. Hypochlorous acid vapour was included to control diseases during hypobaric storage. Exposure to an air flow which had been in contact with hypochlorous acid at 0.25, 1.5 or 5.25 % had no mould after 60 days at 1.7–4.4 °C. The effectiveness of short hypobaric exposure against postharvest rots had also been investigated by Romanazzi et al. (2001). They found that exposure of bunches of the cultivar Italia to 190 mm Hg for 24 h significantly reduced the incidence of grey mould during subsequent storage. Grapes were also wounded and inoculated after hypobaric treatment and it was determined that this treatment decreased infection and diameter of lesions significantly compared to the untreated fruits.

Grapefruits

Haard and Salunkhe (1975) mentioned two reports that suggested that hypobaric storage could extent the postharvest life of grapefruit compared to cold storage alone. In one report the increase was from 20 days at normal atmospheric pressure to 3–4 months under hypobaric conditions and in the other from 30 to 40 days at normal atmospheric pressure to 90–120 days under hypobaric storage. Hypobaric storage at 380 mm (the lower limit of the experimental equipment) and 4.5 °C had no effect on the incidence of chilling injury (Grierson 1971).

Growing Plants

Hypobaric conditions have also been investigated for their effects on growing crops. Part of the justification for this work is related to achieving plant production in a closed ecological life support system (CELSS) in space. Corey et al. (1996) studied the growth of lettuce plants in a controlled ecological life support system that could involve the use of hypobaric pressures to enable lower mass requirements for atmospheres and possible enhancement of crop productivity. When the pressure in a controlled environment growth chamber was reduced from ambient to 388 mm Hg, the rate of net photosynthesis of the plants increased by 25 % and the dark respiration rate decreased by 40 %. In their review, Ishigami and Goto (2008) also reported that photosynthesis and transpiration of plants was higher under hypobaric pressures because gas diffusion rates increased. They gave examples of vegetative growth of spinach and lettuce being normal under 190–380 mm Hg. Rice and Arabidopsis thaliana seeds germinated at 190 mm Hg and seed growth of soybean and A. thaliana under hypobaric conditions was greater than under atmospheric pressure with the same O_2 partial pressure. Flowering of A. thaliana was normal under hypobaric conditions. He and Davies (2012) grew lettuce under 190 mm Hg or atmospheric pressure for 32 days. Conversely they found that significant levels of endogenous ethylene occurred by the 15th day resulting in reductions in photosynthesis, dark respiration rate and a subsequent decrease in plant growth. Hypobaric conditions did not mitigate the adverse ethylene effects on plant growth. Seed germination was not adversely affected by hypobaric conditions and was higher under 190 mm Hg than atmospheric pressure. Previously He et al. (2003) had shown reduced ethylene biosynthesis in wheat and lettuce plants by up to 65 % while increasing plant growth.

Hypobaric storage of growing pot plants and cuttings has been evaluated to ensure maintenance of their quality during storage and distribution. For example, Burg (1973, 2004) described successful storage of chrysanthemum cutting at 0–4 °C under 100–152 mm Hg and at 0–2 °C under 61 mm Hg. Rooted cuttings were also successfully stored at 3.3 °C under 61–81 mm Hg and 0–2 °C under 41–61 mm Hg for more than 84 days. Burg (2004) also reported on storage tests for several species of potted plants at 16.7 °C and 95 % r.h. for 21 days and found that the optimum hypobaric conditions were 30 mm Hg. Andersen and Kirk (1986) reported higher water loss for Hibiscus cuttings stored under hypobaric conditions than cuttings stored under normal atmospheric pressure. They found that when the cuttings are stored under hypobaric conditions the stomata opened, which they explained was probably caused by the low partial pressure of CO_2 in the container since the sensitivity of stomata to CO_2 depended on the hormonal balance of the leaves.

Kohlrabi

Bangers (1974) reported better leaf retention in kohlrabi in storage at 2–4 °C under 75 mm Hg compared with those stored under atmospheric pressure at the same temperature.

Leeks

Ward (1975) found no improvement in storage of leeks under 76 mm Hg and high humidity.

Lettuce

It was claimed that hypobaric storage increased the storage life of lettuce from 14 days in conventional cold stores up to 40–50 days (Haard and Salunkhe 1975). However, Ward (1975) found no improvement in storage under 76 mm Hg and high humidity and Bangers (1973) found no improvement in storage under 75 mm Hg. Burg (2004) reported an increase in the physiological disorder, pink rib, when they were exposed to 30 or 80 mm Hg at 0–1 °C for 4 weeks. Conversely Jamieson (1980 quoted by Burg 2004) found that storage at 0.5 °C and 90–95 % r.h. under 10, 20, 40 or 80 mm Hg resulted in the lettuce remaining in excellent condition after 63 days storage. The lettuce that had been stored in the same conditions at atmospheric pressure had significant incidence of pink rib, black heart and decay after only 37 days storage. Burg (2004) also reported that the manufacturer of hypobaric storage containers, Grumman Allied Industries, investigated marginal and pink discoloration and found that during storage at 5 mm Hg over 35 days the disorder was "eliminated". However, it was "progressively accentuated" at 10–40 mm Hg and "then decreased in frequency as the pressure was increased from 80 to 160 mm Hg". It was concluded that the best storage conditions was 2 °C and 5 mm Hg for 21 days but when the lettuce were stored for 36 days there was a high incidence of russet spotting. An et al. (2009) studied packaging of curled lettuce in small rigid containers at different hypobaric conditions. In both 190 and 380 mm Hg the lettuce had high moisture loss without any observable benefit in keeping quality compared with atmospheric pressure storage.

Limes

Spalding and Reeder (1976) reported that limes coated with wax containing 0.1 % of either of the fungicides thiabendazole or benomyl remained green and suitable for marketing after 3–4 weeks under hypobaric storage of 170 mm Hg at 21.1 °C. They also found that hypobaric storage did not affect chilling injury, but Pantastico (1975) reported that storage in 7 % O_2 reduced the symptoms of chilling injury compared to storage in air. Haard and Salunkhe (1975) stated that Tahiti limes could be stored for 14–35 days in cold storage, but this was extended to 60–90 when it was combined with hypobaric conditions. Spalding and Reeder (1974) stored Tahiti limes under 152 mm Hg pressure for 6 weeks at 10 °C. They showed only small changes in colour, rind thickness, juice content, total soluble solids, total acids and ascorbic acid. Decay averaged 7.8 % compared to those stored in air that was 8.3 %. Those stored at 228 mm Hg maintained acceptable green colour, but were a slightly lighter green than limes at 152 mm Hg. Those stored under 76 mm Hg maintained acceptable green colour, but had low juice content, thick rinds and a high incidence of decay. Limes from all treatments had acceptable flavour. In subsequent work Spalding and Reeder (1976) found that Tahiti limes retained green colour, juice content and flavour acceptable for marketing and had a low incidence of decay during storage at a pressure of 170 mm Hg for up to 6 weeks at 10 or 15.6 °C with 98–100 % r.h. Fruits stored under atmospheric pressure turned yellow within 3 weeks. They concluded that hypobaric storage, but not controlled atmosphere storage, could be used to extend the storage life of limes, which may provide an advantage for hypobaric storage. They also reported higher weight loss and shrivelling of limes during storage under hypobaric conditions compared to those stored at atmospheric pressure.

Loquat

Hypobaric storage at 40–50 mm Hg reduced decay by 87 % and also reduced fruit browning, flesh "leatheriness", respiration rate and ethylene production of loquat fruits compared to those stored under atmospheric pressure at 2–4 °C for 49 days. Peroxidase and polyphenol oxidase activities increased and then reached the highest values after 14 days in air storage. Hypobaric storage reduced the increase in polyphenol oxidase and pyrogallol peroxidase activity compared to those under atmospheric pressure storage and delayed the onset of their peak activity (Gao et al. 2006).

Jujube

Considerable work has been reported from China on the hypobaric storage of jujube (Zizyphus jujuba). Chang Yan Ping (2001) stored the cultivars LiZao and DongZao under hypobaric conditions. They found that during storage the decrease of firmness and the rate of browning were both significantly (p = 0.01) inhibited in storage under 154 mm Hg compared to those under hypobaric conditions. Weight loss under hypobaric storage was less than 2.5 %. It was also reported that the fruits turned red slower under 154, 385 and 616 mm Hg. Wang et al. (2007) that storage of DongZao soften more slowly and had better retention of ascorbic acid and other organic acids than those stored at normal pressure. Cao Zhi Min (2005) found that hypobaric storage significantly retained firmness and vitamin C content, reduced acetaldehyde and ethanol contents in pulp and respiratory rate, inhibited ascorbic acid oxidase and alcohol dehydrogenase activities, slowed down the rate of ethylene production in fruit. Hypobaric storage significantly reduced "number of climacteric peak", but did not postpone climacteric peak (sic). The work was on the cultivars Huanghua, Zhanhual, Zhanhua 2, Shandong Wudi and Dagang harvested different stages of maturity. Cui (2008) reported that hypobaric storage, under 20.3, 50.7 or 101.3 kPa, decreased the respiration rate, delayed loss of ascorbic acid, decreased the rate of superoxide anion production, and prolonged the postharvest life of the cultivar Lizao. There were no significant differences between 20.3 kPa and 50.7, therefore 50.7 kPa was recommended to reduce the cost. Jin et al. (2006) compared storage in different hypobaric conditions and found 55.7 kPa resulted in retention of the best quality.

Mango

Ilangantileke (1989) tested storage of the cultivar Okrang at 13 °C and 60–100 mm Hg and found that they kept well for 4 weeks. Apelbaum et al. (1977b) observed that at pressures below 0.05 MPa (50 mm Hg) mangoes underwent desiccation and, even upon ripening, did not develop their natural red and orange colour. However, fruits submitted to pressures of 0.013 and 0.01 MPa (10 and 13 mm Hg) had an extended storage life of 25 and 35 days respectively. Several Florida varieties of mango that were ripened in air at ambient temperatures showed less anthracnose (*C. gloeosporioides*) and stem-end rot (*Diplodia natalensis*) when they had previously been stored hypobaric conditions at 13 °C. The reduction in decay coincided with a retardation in fruit ripening, permitting a prolonged storage at 13 °C (Spalding and Reeder 1977). It is possible that this effect on anthracnose and stem-end rot may have been a reflection of the delay in ripening since the development of postharvest diseases of mangoes, particularly anthracnose, develops as the fruit ripens (Thompson 2015).

Oat Leaves

Veierskov and Kirk (1986) demonstrated a significant decrease in the respiration rate of sections of oat leaves stored under hypobaric conditions compared to those stored under normal atmospheric pressures.

Okra

Knee and Aggarwal (2000) found okra showed less darkening of seeds in vacuum containers at 380 mm Hg than containers without a vacuum.

Onions

McKeown and Lougheed (1981), Hardenburg et al. (1990) and Burg (2004) reported that bulb onions stored at 26 °C and 58 % r.h. under 61 mm Hg for

28 days lost 12.2 % in weight. McKeown and Lougheed 1981) reported that there was potential for the use of hypobaric conditions of 61 mm Hg in 26 °C and low humidity for the curing of onions.

Burg (2004) found that at 0–3 °C spring onions could be stored for almost 3 weeks under 50–60 mm Hg compared to less than 6 days under atmospheric pressure. Also that storage under 100–150 mm Hg had only a slight beneficial effect compared to atmospheric pressure. Storage at 1 °C under 61 mm Hg and near-saturation humidity had little effect on the content of leaf chlorophyll of spring onions after 21 days. In general, their weight loss after hypobaric storage or the controlled atmosphere storage under 2 % O_2 + 0 % CO_2 was less than that which occurred after air storage. The spring onions also retained a fresh appearance after hypobaric storage but there was no difference among treatments in the sensory evaluation (McKeown and Lougheed 1981). They also found better chlorophyll retention in storage at 1 °C under 55–60 mm Hg than under atmospheric pressure or controlled atmosphere storage. However, Ward (1975) found no improvement in storage under 76 mm Hg and high humidity.

Oranges

Min and Oogaki (1986) reported a reduction in the respiration rate of oranges stored under hypobaric conditions compared to those stored under atmospheric pressure. Their respiration rate remained lower even after the fruit were transferred to atmospheric pressure. The orange cultivar Fukuhara had lower rates of ethylene production and respiration rate after hypobaric storage compared to those fruits under atmospheric pressure throughout. Oranges stored at 190 mm Hg and 98 % r.h. decayed more rapidly after return to atmospheric pressure than fruits stored at 190 mm Hg and 75 % r.h. or atmospheric pressure and 86 % r.h. throughout (Min and Oogaki 1986). However, the fruits stored at these humidity levels and low pressure would probably have been severely desiccated.

Papayas

Papaya stored at 10 °C and 98 % r.h. and a pressure of 20 mm Hg ripened more slowly and had less disease development than fruit at atmospheric pressure in refrigerated containers (Alvarez 1980). Alvarez and Nishijima (1987) also reported that hypobaric storage appeared to suppress postharvest disease development in papaya fruit.

Parsley

It was reported that storage in at 3 °C under hypobaric conditions of 75 mm Hg extended the storage life of parsley from 5 weeks in atmospheric pressure to 8 weeks without appreciable losses in protein, ascorbic acid or chlorophyll content (Bangerth 1974).

Peaches

Hypobaric storage prolonged the postharvest life of peaches from 7 days at atmospheric pressure to 27 days. It delayed chlorophyll and starch breakdown, carotenoid synthesis and the decrease in sugars and acidity (Salunkhe and Wu 1973). Storage at 11–12 °C in an aerated hypobaric system at either 190 or 76 mm Hg retarded ripening compared to storage at the same temperature under atmospheric pressure (Kondou et al. 1983). In the cultivar Okubo, that had been stored under hypobaric conditions of 108 mm Hg, the ripening rate after storage was slower than that of fruits that had been stored in air at atmospheric pressure throughout. The physiological disorder, mealy breakdown, was reduced under hypobaric storage, but no effect was found on flesh browning or on abnormal peeling (Kajiura 1975). Porritt S.W. and Woodruff R.E. (quoted by Pattie and Lougheed 1974) suggested "that LPS (low pressure storage) has not proven suitable for peaches". They recommended that additional research was needed to assess further the value of hypobaric storage to extend storage life of peaches. Jinhua Wang et al. (2015) stored Honey peach (Prunus persica) under 101, 10–20, 40–50 or 70–80 kPa for 30 days at 0 °C and at 85–90 % r.h followed by 4 days at 25 °C and 80–85 % r.h. They found that 10–20 kPa delayed decay rates, maintained overall quality and extended the shelf life. They showed that it effectively delayed increases in both O_2 radical dot- production rate and ozone content, enhanced activities of catalase and superoxide dismutase and increased contents of adenosine triphosphate and adenosine diphosphate while reducing lipoxygenase activity and adenosine monophosphate content during the storage period as well as the following shelf life.

Pears

Hypobaric conditions prolonged the storage life of pears by between 1.5 and 4.5 months compared to cold storage alone. It delayed chlorophyll and starch breakdown, carotenoid formation and the decrease in sugars and acidity (Salunkhe and Wu 1973).

Pineapples

Staby (1976) stated that storage of pineapple under hypobaric condition can extend the storage life for up to 30–40 days.

Plums

Prunes (*Prunus domestica* cultivars that are grown mainly for dried fruit production; prune is the French word for plum) were successfully stored under 40–101 mm Hg by Patterson and Melsted (1978).

Potatoes

Greening and glycoalkaloids synthesis, including solanine, can occur in stored potato tubers when they are exposed to strong light (Mori and Kozukue 1995). Burg (2004) reported that greening in light was inhibited in potato tubers stored at 15 °C and 126 mm Hg but there was no effect on their solanine content. He also reported that similar results were also reported during storage at 2.5, 5, 10 or 20 °C and 76–95 mm Hg and under controlled atmosphere storage with 2–2.5 % O_2.

Radish

McKeown and Lougheed (1981) reported that during storage at 1 °C under 55–60 mm Hg for 7–28 days the weight losses were similar to the levels reported for controlled atmosphere storage with 2 % O_2 and slaked lime to control CO_2. However, off-flavours developed in the controlled atmosphere stored radishes but not in those under hypobaric storage. Bangerth (1974) found that in storage at 2.2–2.8 °C and 95 % r.h. there was decreased losses of protein, ascorbic acid and chlorophyll losses under 75 mm Hg compared to storage in the same conditions under atmospheric pressure. In general, the weight loss during storage at 1 °C under 61 mm Hg and near-saturation humidity and controlled atmosphere storage in 2 % O_2 + 0 % CO_2 was less than that which occurred during storage under atmospheric pressure. Radishes also retained a fresh appearance after storage under 61 mm Hg and their sensory evaluation indicated that they were similar to those held in air, while those held under 2 % O_2 + 0 % CO_2 atmosphere had a lower rating and off-flavours (McKeown and Lougheed 1981). However, Ward (1975) found no improvement in storage under 76 mm Hg and high humidity.

Spinach

Bangerth (1973) found that storage of spinach at 3 °C and 95 % r.h. with 75 mm Hg retained their green colour, vitamin C and total protein content for almost seven weeks.

Squash

Burg (2004) reported that storage of Yellow Crookneck squash at 7.2 °C and 65, 80 or 150 mm Hg had no beneficial effects compared to storage at atmospheric pressure. McKeown and Lougheed (1981) carried out limited trials with Acorn squash at 10 °C and at low humidity under 61 mm Hg for 42 days after curing, which "showed some potential".

Strawberries

It was reported by Haard and Salunkhe (1975) that the strawberry cultivars Tioga and Florida 90 in cold storage could be stored for 21 days under hypobaric pressure compared to only 5–7 °C at atmospheric pressure. Bubb (1975a) compared storage of the cultivar Cambridge Favourite at 3.3 °C under atmospheric pressure with storage under 25 and 101 mm Hg and found better flavour retention under hypobaric storage. When he compared this with storage in O_2 partial pressures similar to those obtained under the hypobaric conditions he did not observe similar flavour retention. There was some evidence of slight tainting of the flavour after hypobaric storage but after subsequent shelf life at 18.5 °C he reported that the tainting of the flavour of the fruit became objectionable. After 10 days storage none of the samples exceeded 2 % weight loss but there was high levels of rotting in all samples but no consistent effect on fruit firmness. Knee and Aggarwal (2000) found a higher proportion of strawberry fruits were infected by fungi in vacuum containers at 380 mm Hg than in containers with no vacuum. An et al. (2009) also studied hypobaric packaging in rigid small containers during storage at 3 °C with intermittent fresh air flushing and vacuum treatment to avoid creating an anoxic environment. Bacterial growth was slightly reduced in fruit in hypobaric packaging but there was an operational drawback with the 25.3 kPa conditions in that fresh air flushing and repetitive vacuum application were required too frequently for the densely packed strawberry packages.

The effectiveness of short hypobaric treatments against postharvest rots was investigated by Romanazzi et al. (2001). On Pajaro, the greatest reductions of *B. cinerea* and *Rhizopus stolonifer* rot were observed on fruits treated for 4 h at 0.25 and 0.50 atmospheres, respectively. Hashmi et al. (2013a) exposed strawberries to hypobaric pressures of 190, 380 and 570 mm Hg for 4 h at 20 °C and subsequently stored them at 20 or 5 °C. They found that exposure to 380 mm Hg consistently delayed rot development both at 20 and 5 °C and did not affect their weight loss and firmness, but an initial increase in respiration rate was observed. They also found (Hashmi et al. 2013b) that exposure of strawberries to 380 mm Hg for 4 h reduced rot incidence from natural infection during subsequent storage for 4 days at 20 °C. The same treatment also had the same effect when the fruit had been inoculated with *B. cinerea* or *R. stolonifer* spores. They found that activities of defence-related enzymes were increased after exposure to hypobaric conditions. PAL (EC: 4.3.1.24) and chitinase (EC: 3.2.1.14) peaked 12 h after exposure, while POD (EC: 1.11.1.7) increased immediately after exposure. PPO (EC: 1.10.3.1) activity remained unaffected during subsequent storage for 48 h at 20 °C. However they found that exposure to atmospheric pressure for 4 h did not influence rot development.

Sweetcorn

Haard and Salunkhe (1975) found that hypobaric storage of sweetcorn could increase their storage life to 21 days from only 4–8 days in a conventional cold storage at atmospheric pressure. Burg (2004) stored the cultivar Wintergreen at 1.7 ± 1 °C at pressures over the range of 10–760 mm Hg and found that their respiration rate decreased with decreasing pressure. After 7 days a sensory evaluation panel determined that those stored under 20 mm Hg had the highest scores for sweetness and flavour, while after 11 days those that had been stored under 50 mm Hg had the best flavour score.

Tomatoes

Storage of mature green tomatoes under hypobaric conditions of 646, 471, 278 or 102 mm Hg at 12.7 °C resulted in a reduction in fruit respiration, especially under 102 mm Hg (Wu and Salunkhe 1972). Wu et al. (1972) stored tomatoes for 100 days under 102 mm Hg and then transferred them to 646 mm Hg, both at 12.8 °C and 90–95 % r.h., where they ripened normally based on their lycopene, chlorophyll, starch, sugar and β-carotene content. However, their aroma was inferior to those that had been stored under atmospheric pressure. Xiaoqing Dong et al. (2013) demonstrated that mid-climacteric tomatoes harvested at the breaker or pink stages exposed to 76 mm Hg for 6 h showed transient increased sensitivity to 1-MCP. Subsequently they found that mid-climacteric fruit exposed to 500 nl L − 1 1-MCP under 76 or 16 mm Hg for 1 h showed acute disturbance of ripening. They concluded that high efficacy of 1-MCP applied under hypobaric conditions was due to rapid ingress and accumulation of internal gaseous 1-MCP.

Turnips

Onoda et al. (1989) found that turnips retained better appearance and lower weight loss when stored in cycles of hypobaric pressures (between 100 and 300 mm Hg with no humidification) than those stored at a constant atmospheric pressure.

Watercress

Hypobaric storage of watercress at 3 °C and 75 mm Hg for 12 days resulted in them retaining their ascorbic acid content better than those stored under atmospheric pressure. There was also better retention of protein and chlorophyll content under hypobaric storage (Bangerth 1974).

Enhancement

As with any other technique, storage under hypobaric conditions has been successfully used in combination with other treatments. For example Romanazzi et al. (2003) found that the combination of spraying with chitosan at 0.1, 0.5 or 1.0 % 7 days before harvest and exposure to 380 mm Hg for 4 h directly after harvest effectively controlled fungal decay of sweet cherries during 14 days storage at 0 ± 1 °C, followed by a 7 day shelf life. Fungi associated with rots included brown rot (*Monilinia sp.*), grey mould (*B. cinerea*), blue mould (*P. expansum*), Alternaria rot, (*Alternaria* sp.) and Rhizopus rot (*Rhizopus* sp.). Yoshiki Kashimura et al. (2010) found that applying 1-MCP to Jonagold and Fuji apples and Shinsei and Shinsui Japanese pears under an atmosphere of 152 mm Hg reduced the exposure time required to have the same effect as applying 1-MCP at atmospheric pressure. Xiaoqing Dong et al. (2013) concluded that high efficacy of 1-MCP applied to tomatoes under hypobaric conditions was due to rapid ingress and accumulation of the gaseous 1-MCP. They had found that mid-climacteric tomatoes exposed to 500 nl L − 1 1-MCP under 76 or 16 mm Hg for 1 h showed acute disturbance of ripening. Spalding and Reeder (1976) found that limes coated with wax containing 0.1 % of either of the fungicides thiabendazole or benomyl remained green and suitable for marketing after 3–4 weeks under hypobaric storage of 170 mm Hg at 21.1 °C.

Burg (2013 personal communication) described a method for generating hypochlorous acid vapour into a commodity packed for shipment in a hypobaric intermodal container. Within 1 h the vapour was 100 % effective in killing bacteria, fungi and viruses both on the plant surfaces and within its interior, without injuring the commodity. The Food and Drug Administration of the USA were reported to have given clearance for use of hypochlorous acid vapour in a hypobaric system and Burg has filed a new patent on hypobaric storage.

Vacuum Infiltration

Vacuum infiltration is where fruit or vegetables are placed in a liquid inside a container and a vacuum is applied to the container that sucks out air and other gases from their intercellular spaces. When the vacuum is released the liquid fills these vacated air spaces. It is a method that has been used to ensure chemicals that may improve the postharvest life of the fruit or vegetables reaches more of the cells. For example vacuum infiltration of shiitake mushrooms with calcium chloride solution markedly inhibited respiration rate and ethylene production, decreased the loss of total soluble solids, protein, reducing sugars, starch, organic acids, ascorbic acid and fibre, significantly restricted the increase of cell membrane permeability and thus retained their quality of during storage (Li et al. 2000). Haruenkit and Thompson (1996) showed similar results with vacuum infiltration of calcium into pineapple fruits. Vacuum infiltration of calcium has also been shown to inhibit ripening of tomatoes (Wills and Tirmazi 1979), avocados (Wills and Tirmazi 1982) and mangoes (Wills and Tirmazi 1981). These treatments were found significantly to delay ripening without affecting fruit quality. In these three latter cases the calcium was applied by vacuum infiltration under 250 mm Hg with calcium chloride at concentrations within the range of 1 to 4 %. They found that higher concentrations could result in skin damage. Vacuum (32 kPa) and pressure infiltration (115 kPa) of Kensington Pride mangoes with 2–8 % calcium chloride solution resulted in delayed softening of fruit during storage at 20 °C of 8–12 days compared to untreated fruit (Yuen 1993). However, peel colour remained partially green when fruits were ripe and some peel injury occurred in treated fruit. Cling film (19 μm thick), shrink film (17 μm thick) of polyethylene bags (50 μm thick) appeared to be as effective as calcium infiltration in delaying ripening without peel injury and undesirable retention of excessive green colour in the ripe fruit (Yuen 1993). Drake and Spayd (1983) vacuum infiltrated Golden Delicious apples with calcium chloride (3 % at 3 pounds per square inch for 8 min) before storing them for 5 months at 1 °C. These fruit were firmer and more acid than untreated fruit stored for the same period. Yuen (1993) concluded that calcium infiltration was shown reduce chilling injury and increase disease resistance during storage.

Vacuum Cooling

Cooling of horticultural crops may be achieved by using the latent heat of vaporisation of the water within the crop. To achieve this, the crop is placed in a vacuum chamber at reduced pressure, which increases water evaporation from their surfaces. This, in turn, extracts heat from the crop to provide the energy to vaporise the water. For every 5 or 6 °C reduction in temperature, Barger (1963) found that the crop would lose some 1 % in weight. Vacuum coolers have to be strongly constructed from heavy duty steel and are usually cylindrical in shape to withstand the low pressure without imploding. The vacuum is usually achieved by a vacuum pump attached to the cylinder. The speed and effectiveness of cooling is related to the ratio between the mass of the crop and its surface area so it is particularly suitable for leaf crops such as lettuce. Where there is a low ratio between mass and surface area or there is an effective barrier to water loss from the crop surface, vacuum cooling can be unacceptably slow. For example, tomatoes have a low ratio between mass and surface area and a relatively thick wax cuticle and are therefore not suitable for vacuum cooling.

References

Adams, K.B., M.T. Wu, and D.K. Salunkhe. 1976. Effects of subatmospheric pressure on the growth and patulin production of *P. expansum* and *P. patulum*. *LWT—Food Science and Technology* 9: 153–155.

Aharoni, Y., and L.G. Houck. 1980. Improvement of internal color of oranges stored in oxygen-enriched atmospheres. *Scientific Horticulture* 13: 331–338.

Aharoni, Y., A. Apelbaum, and A. Copel. 1986. Use of reduced atmospheric pressure for control of the green peach aphid on harvested head lettuce. *HortScience* 21: 469–470.

Al-Qurashi, A.D., F.B. Matta, and J.O. Garner. 2005. Effect of using low-pressure storage (LPS) on Rabbiteye blueberry 'Premier' fruits. *Journal of King Abdulaziz University: Meteorology, Environment and Arid Land Agriculture* 16: 3–14.

Alvarez, A.M. 1980. Improved marketability of fresh papaya by shipment in hypobaric containers. *HortScience* 15: 517–518.

Alvarez, A.M., and W.T. Nishijima. 1987. Postharvest diseases of papaya. *Plant Diseases* 71: 681–686.

Andersen, A.S., and H.G. Kirk. 1986. Influence of low pressure storage on stomatal opening and rooting of cuttings. *Acta Horticulturae* 181: 305–311.

Anonymous (undated). A broad look at the practical and technical aspects of hypobaric storage. Dormavac Corporation, Miami, Florida, USA.

Anonymous. 1975. Annotated bibliography on low pressure (Hypobaric) storage, 1968–1973. *Commonwealth Bureau of Horticulture and Plantation Crops: East Malling* 6326.

Apelbaum, A., Y. Aharoni, and N. Tempkin-Gorodeiski. 1977a. Effects of subatmospheric pressure on the ripening processes of banana fruits. *Tropical Agriculture* 54: 39–46.

Apelbaum, A., and R. Barkai-Golan. 1977. Spore germination and mycelia growth of postharvest pathogens under hypobaric pressure. *Phytopathology* 67: 400–403.

Apelbaum, A., G. Zauberman, and Y. Fuchs. 1977b. Subatmospheric pressure storage of mango fruits. *Scientia Horticulturae* 7: 153–160.

Back, E.A., and R.T. Cotton. 1925. The use of vacuum for insect control. *Journal of Agriculture Research* 31: 1035–1041.

Bangerth, F. 1973. The effect of hypobaric storage on quality, physiology and storage life of fruits, vegetables and cut flowers. *Gartenbauwissenschaft* 38: 479–508.

Bangerth, F. 1974. Hypobaric storage of vegetables. *Acta Horticulturae* 38: 23–32.

Bangerth, F. 1984. Changes in sensitivity for ethylene during storage of apple and banana fruits under hypobaric conditions. *Scientia Horticulturae* 24: 151–163.

Bangerth, F., and J. Streif. 1987. Effect of aminoethoxyvinylglycine and low-pressure storage on the post-storage production of aroma volatiles by golden delicious apples. *Journal of the Science of Food and Agriculture* 41: 351–360.

Bare, C.O. 1948. The effect of prolonged exposure to high vacuum on stored tobacco insects. *Journal of Economic Entomology* 41: 109–110.

Barger, W.R. 1963. Vacuum precooling. A comparison of cooling of different vegetables. *USA Department of Agriculture, Market Research Report No.* 600.

Ben-Yehoshua, S., and V. Rodov. 2003. Transpiration and water stress. In *Postharvest physiology and pathology of vegetables*, ed. J.A. Bartz, and J.K. Brecht, 111–159. New York: Marcel Dekker.

Berrios, J.D.J., B.G. Swanson, and W.A. Cheong. 1999. Physico-chemical characterization of stored black beans (*Phaseolus vulgaris* L.). *Food Research International* 32: 669–676.

Blackbourn, H.D., M.J. Jeger, P. John, and A.K. Thompson. 1990. Inhibition of degreening in the peel of bananas ripened at tropical temperatures, III changes in plastid ultrastructure and chlorophyll protein complexes accompanying ripening in bananas and plantains. *Annals of Applied Biology* 117: 147–161.

Borecka, H.W., and K. Pliszka. 1985. Quality of blueberry fruits (*Vaccinium corymbosum* L.) stored under LPS, CA and normal air storage. *Acta Horticulturae* 165: 241–246.

Bosland, P.W., and E.J. Votava. 2000. *Peppers, vegetables and spices*. New York: CABI Publishing.

Brednose, N. 1980. Effects of low pressure on storage life and subsequent keeping quality of cut roses. *Acta Horticulturae* 113: 73–79.

Brown, W. 1922. On the germination and growth of fungi at various concentrations of oxygen and carbon dioxide. *Annals of Botany* 36: 257–283.

Bubb, M. 1975a. Hypobaric storage. Annual report of the East Malling Research Station, UK. for 1974 p. 83.

Bubb, M. 1975b. Hypobaric storage. Annual report of the East Malling Research Station, UK. for 1974 p. 81.

Bubb, M. 1975c. Hypobaric storage. Annual report of the East Malling Research Station, UK. for 1974 pp. 77–78.

Bubb, M., and I.W. Langridge. 1974. Low pressure storage. Annual report of the East Malling Research Station, UK. for 1973 p. 104.

Burdon, J., N. Lallu, G. Haynes, K. McDermott, and D. Billing. 2008. The effect of delays in establishment of a static or dynamic controlled atmosphere on the quality of 'Hass' avocado fruit. *Postharvest Biology and Technology* 49: 61–68.

Burg, S.P. 2014. *Hypobaric storage in food industry*. London: Academic Press.

Burg, S.P., and E.A. Burg. 1965. Ethylene action and the ripening of fruits. *Science* 148: 1190–1196.

Burg, S.P., and E.A. Burg. 1966a. Fruit storage at subatmospheric pressure. *Science* 153: 314–315.

Burg, S.P., and E.A. Burg. 1966b. Relationship between ethylene production and ripening of bananas. *Botanical Gazette* 126: 200–204.

Burg, S.P. 1967. Method for storing fruit. US Patent3.333967 and US patent reissue Re. 28,995 (1976).

Burg, S.P. 1973. Hypobaric storage of cut flowers. *HortScience* 8: 202–205.

Burg, S.P. 1975. Hypobaric storage and transportation of fresh fruits and vegetables. In *Postharvest biology and handling of fruits and vegetables*, ed. N.F. Haard, and D.K. Salunkhe, 172–188. Westpoint: A.V.I. Publishing Company Inc.

Burg, S.P. 1990. Theory and practice of hypobaric storage. In *Food preservation by modified atmospheres*, ed. M. Calderon, and R. Barkai-Golan, 353–372. Boca Raton: CRC Press.

Burg, S.P. 1993. Current status of hypobaric storage. *Acta Horticulturae* 326: 259–266.

Burg, S.P. 2004. *Postharvest physiology and hypobaric storage of fresh produce*. Wallingford: CAB International.

Burg, S.P. 2010. Experimental errors in hypobaric laboratory research. *Acta Horticulturae* 857: 45–62.

Burg, S.P., and R. Kosson. 1983. Metabolism, heat transfer and water loss under hypobaric conditions. In *Postharvest physiology and crop preservation*, ed. M. Lieberman, 399–424. New York: Plenum Corp.

Burton, W.G. 1989. *The potato*, 3rd ed. London: Longmans.

Caldwell, J. 1965. Effects of high partial pressures of oxygen on fungi and bacteria. *Nature* 206: 321–323.

Cao Zhi Min 2005. *The study on hypobaric storage mechanism and technology in dong-zao jujube fruit*. Master's thesis, Tianjin University of Science and Technology. http://www.dissertationtopic.net/doc/1089396. Accessed October 2013.

Chang Yan Ping 2001. *Study on the physiological-biochemical changes and storage effects of jujube fruits under hypobaric (low pressure) condition*. Master's thesis, Shanxi Agricultural University. http://www.dissertationtopic.net/doc/813061. Accessed October 2013.

Chau, K.F., and A.M. Alvarez. 1983. Effects of LP storage on Collectotrichum gloeosporioides and postharvest infection of papaya. *HortScience* 18: 953–955.

Chen, Z.J., M.S. White, and W.H. Robinson. 2005. Low-pressure vacuum to control larvae of *Hylotrupes bajulus* (Coleoptera: Cerambycidae). In *Proceedings of the fifth international conference on urban pests*, ed. C.Y. Lee, and W.H. Robinson. Malaysia: Perniagaan Ph'ng @ P&Y Design Network.

Choudhury, J.K. 1939. Researches on plant respiration. v. on the respiration of some storage organs in different oxygen concentrations. *Proceedings of the Royal Society, London Series B* 127: 238–257.

Corey, K.A., M.E. Bates, S.L. Adams, and R.D. MacElroy. 1996. Carbon dioxide exchange of lettuce plants under hypobaric conditions. *Advances in Space Research* 18: 301–308.

Couey, H.M., M.N. Follstad, and M. Uota. 1966. Low oxygen atmospheres for control of post-harvest decay of fresh strawberries. *Phytopathology* 56: 1339–1341.

Cui, Y. 2008. Effects of hypobaric conditions on physiological and biochemical changes of Lizao jujube. *Journal Anhui Agricultural Science* 36: 12900–12901.

Davenport, T.L., S.P. Burg, and T.L. White. 2006. Optimal low pressure conditions for long-term storage of fresh commodities kill Caribbean fruit fly eggs and larvae. *HortTechnology* 16: 98–104.

Dilley, D.R. 1972. Hypobaric storage—a new concept for preservation of perishables. In: *Proceedings of the Michigan State Horticultural Society*, pp. 82–89.

Dilley, D.R. 1977. Application of the hypobaric system for storage and transportation of perishable agricultural commodities. In: *2nd Annual world's fair for technology exchange*, 7–11 February 1977, pp. 135–149.

Dilley, D.R. 1990. Historical aspects and perspectives of controlled atmosphere storage. In *Food preservation by modified atmospheres*, ed. M. Calderon, and R. Barkai Golan, 187–196. Ann Arbor: CRC Press Boca Raton.

Dilley, D.R., and W.J. Carpenter. 1975. Principles and application of hypobaric storage of cut flowers. *Acta Horticulturae* 41: 249–267.

Dilley, D.R., P.L. Irwin, and M.W. McKee. 1982. Low oxygen, hypobaric storage and ethylene scrubbing. In *Controlled atmosphere storage and transport of perishable agricultural commodities*, ed. D.G. Richardson, and M. Meheriuk, 317–329. Beaverton: Timber Press.

Dolt, K.S., J. Karar, M.K. Mishra, J. Salim, R. Kumar, S.K. Grover, and M.A. Qadar Pasha. 2007. Transcriptional down regulation of sterol metabolism genes in murine liver exposed to acute hypobaric hypoxia. *Biochemical and Biophysical Research Communications* 354: 148–153.

Dong, Xiaoqing, D.J. Huber, Jingping Rao, and J.H. Lee. 2013. Rapid ingress of gaseous 1-MCP and acute suppression of ripening following short-term application to mid-climacteric tomato under hypobaria. *Postharvest Biology and Technology* 86: 285–290.

Drake, S.H., and S.E. Spayd. 1983. Influence of calcium treatment on "Golden Delicious" apple quality. *Journal of Food Science* 48: 403–405.

An, Duck Soon, Eunyoung Park, and Dong Sun Lee. 2009. Effect of hypobaric packaging on respiration and quality of strawberry and curled lettuce. *Postharvest Biology and Technology* 52: 78–83.

Enfors, S.O., and G. Molin. 1980. Effect of high concentrations of carbon dioxide on growth rate of *Pseudomonas fragi*, *Bacillus cereus* and *Streptococcus cremoris*. *Journal of Applied Bacteriology* 48: 409–416.

Gao, H.Y., H.J. Chen, W.X. Chen, Y.T. Yang, L.L. Song, Y.M. Jiang, and Y.H. Zheng. 2006. Effect of hypobaric storage on physiological and quality attributes of loquat fruit at low temperature. *Acta Horticulturae* 712: 269–274.

Goszczynska, D.M., and M.R. Ryszard. 1988. Storage of cut flowers. *Horticultural Reviews* 10: 35–62.

Grierson W. 1971. Chilling injury in tropical and subtropical fruits: IV. The role of packaging and waxing in minimizing chilling injury of grapefruit. *Proceedings of the Tropical Region, American Society for Horticultural Science* 15: 76–88.

Haard, N.F., and D.K. Salunkhe, 1975. *Symposium: Postharvest biology and handling of fruits and vegetables*. AVI Publishing Company Incorporated, Westpoint.

Chen, Hangjun, Hailong Yang, Haiyan Gao, Jie Long, Fei Tao, Xiangjun Fang, and Yueming Jiang. 2013a. Effect of hypobaric storage on quality, antioxidant enzyme and antioxidant capability of the Chinese bayberry fruits. *Chemistry Central Journal* 7: 4.

Hardenburg, R.E., A.E. Watada, and C.Y. Wang. 1990. The commercial storage of fruits, vegetables and florist and nursery stocks. *United States Department of Agriculture, Agricultural Research Service, Agriculture Handbook* 66.

Haruenkit, R., and A.K. Thompson. 1996. Effect of O_2 and CO_2 levels on internal browning and composition of pineapples Smooth Cayenne. In: *Proceedings of the International Conference on Tropical Fruits*, Kuala Lumpur, Malaysia, 23–26 July 1996, pp. 343–350.

Hashmi, M.S., A.R. East, J.S. Palmer, and J.A. Heyes. 2013a. Pre-storage hypobaric treatments delay fungal decay of strawberries. *Postharvest Biology and Technology* 77: 75–79.

Hashmi, M.S., A.R. East, J.S. Palmer, and J.A. Heyes. 2013b. Hypobaric treatment stimulates defence-related enzymes in strawberry. *Postharvest Biology and Technology* 85: 77–82.

Hatton, T.T., and W.F. Reeder. 1965. Controlled atmosphere storage of Lula avocados-1965 tests. *Proceedings of the Caribbean Region American Society for Horticultural Science* 9: 152–159.

He, C.J., and F.T. Davies. 2012. Ethylene reduces plant gas exchange and growth of lettuce grown from seed to harvest under hypobaric and ambient total pressure. *Journal of Plant Science* 169: 369–378.

He, C.J., F.T. Davies, R.E. Lacey, M.C. Drew, and D.L. Brown. 2003. Effect of hypobaric conditions on ethylene evolution and growth of lettuce and wheat. *Journal of Plant Physiology* 160: 1341–1350.

Herreid, C.F. 1980. Hypoxia in invertebrates. *Comparative Biochemistry and Physiology* 67: 311–320.

Hesselman, C.W., and H.T. Freebairn. 1969. Rate of ripening of initiated bananas as influenced by oxygen and ethylene. *Journal of the American Society for Horticultural Science* 94: 635–637.

Hughes, P.A., A.K. Thompson, R.A. Plumbley, and G.B. Seymour. 1981. Storage of capsicums (*Capsicum annum* [L.] Sendt.) under controlled atmosphere, modified atmosphere and hypobaric conditions. *Journal of Horticultural Science* 56: 261–265.

Chen, Huiyun, Jiangang Ling, Fenghua Wu, Lingju Zhang, Zhidong Sun, and Huqing Yang. 2013b. Effect of hypobaric storage on flesh lignification, active oxygen metabolism and related enzyme activities in bamboo shoots. *LWT—Food Science and Technology* 51: 190–195.

Ilangantileke, S.G., L.Turla, and R. Chen. 1989. Pre-treatment and hypobaric storage for increased storage life of mango. *Canadian American Society of Agricultural Engineering Paper* 896036.

Ishigami, Y., and E. Goto. 2008. Plant growth under hypobaric conditions. *Journal of Science and High Technology in Agriculture* 20: 228–235.

Jamieson, W. 1980. Use of hypobaric conditions for refrigerated storage of meats, fruits and vegetables. *Food Technology* 34: 64–71.

Jiao, S., J.A. Johnson, J.K. Fellman, D.S. Mattinson, J. Tang, T.L. Davenport, and S. Wang. 2012a. Evaluating the storage environment in hypobaric chambers used for disinfesting fresh fruits. *Biosystems Engineering* 7: 271–279.

Jiao, S., J.A. Johnson, J.K. Fellman, D.S. Mattinson, Juming Tang, T.L. Davenport, and S. Wang. 2012b. Evaluating the storage environment in hypobaric chambers used for disinfesting fresh fruits. *Biosystems Engineering* 30: 1–9.

Jiao, S., J.A. Johnson, J. Tang, D.S. Mattinson, J.K. Fellman, T.L. Davenport, and S. Wang. 2013. Tolerance of codling moth and apple quality associated with low pressure/low temperature treatments. *Postharvest Biology and Technology* 85: 136–140.

Jin, A.X., Y.P. Wang, and Liang, L.S. 2006. Effects of atmospheric pressure on the respiration and softening of DongZao jujube fruit during hypobaric storage. *Journal of the Northwest Forestry University* 21: 143–146.

Wang, Jinhua, Yanli You, Wenxuan Chen, Qingqing Xu, Jie Wang, Yingkun Liu, Lili Song, and Jiasheng Wu. 2015. Optimal hypobaric treatment delays ripening of honey peach fruit via increasing endogenous energy status and enhancing antioxidant defence systems during storage. *Postharvest Biology and Technology* 101: 1–9.

Joanny, P., J. Steinberg, P. Robach, J.P. Richalet, C. Gortan, B. Gardette, and Y. Jammes. 2001. Operation Everest III (Comex'97): the effect of simulated sever hypobaric hypoxia on lipid peroxidation and antioxidant defence systems in human blood at rest and after maximal exercise. *Resuscitation* 49: 307–314.

Johnson, J.A., and J.L. Zettler. 2009. Response of postharvest tree nut lepidopoteran pests to vacuum treatments. *Journal of Economic Entomology* 102: 2003–2010.

Kader, A.A. 1989. A summary of CA requirements and recommendations for fruit other than pome fruits. In: *5th International controlled atmosphere conference proceedings*, June 14–16 1989, Wenatchee, Washington, United States of America, Vol. 2, Other commodities and storage recommendations, pp. 303–328.

Kajiura, I. 1975. CA storage and hypobaric storage of white peach 'Okubo'. *Scientia Horticulturae* 3: 179–187.

Kashimura, Yoshiki, Hiroko Hayama, and Akiko Ito. 2010. Infiltration of 1-methylcyclopropene under low pressure can reduce the treatment time required to maintain apple and Japanese pear quality during storage. *Postharvest Biology and Technology* 57: 14–18.

Kidd, F. and C. West. 1927. A relation between the concentration of O_2 and CO_2 in the atmosphere, rate of respiration, and the length of storage of apples. *Report of the Food Investigation Board London for 1925, 1926*, pp. 41–42.

Kidd, F. and C. West, 1934. Injurious effects of atmospheres of pure O_2 upon apples and pears at low temperatures. *Report of the Food Investigation Board, London, UK, for 1933*, pp. 74–77.

Knee, M., and D. Aggarwal. 2000. Evaluation of vacuum containers for consumer storage of fruits and vegetables. *Postharvest Biology and Technology* 19: 55–60.

Kondou, S., C. Oogaki, and K. Mim. 1983. Effects of low pressure storage on fruit quality. *Journal of the Japanese Society for Horticultural Science* 52: 180–188.

Kopec, K. 1980. Zmeny plodov zeleninovej papriky pocas hypobarickeho uskladnovania. *Vedecke Prace Vyskumneho a Slachtitelskeho Ustavu Zeleniny a Specialnych Plodin v Hurbanove* 1: 34–41.

Lafuente, M.T., M. Cantwell, S.F. Yang, and U. Rubatzky. 1989. Isocoumarin content of carrots as influenced by ethylene concentration, storage temperature and stress conditions. *Acta Horticulturae* 258: 523–534.

Langridge, I.W., and R.O. Sharples. 1972. Storage under reduced atmospheric pressure. *Annual Report of the East Malling Research Station, UK. for 1971*, 76.

Laugheed, E.C., D.P. Murr, and L. Berard. 1978. Low pressure storage for horticultural crops. *HortScience* 13: 21–27.

Laurin, É., M.C.N. Nunes, J.-P. Émond, and J.K. Brecht. 2006. Residual effect of low-pressure stress during simulated air transport on Beit Alpha-type cucumbers: Stomata behaviour. *Postharvest Biology and Technology* 41: 121–127.

Li, J.Y., W.N. Huang, L.X. Cai, and W.J. Hu. 2000. Effects of calcium treatment on physiological and biochemical changes in shiitake *Lentinus edodes* during the post harvest period. *Fujian Journal of Agricultural Sciences* 15: 43–47.

Li, W.X. 2006. *Studies on the technology and mechanism of green asparagus during three-stage hypobaric storage*. PhD thesis, Jiangnan University [abstract]. http://www.dissertationtopic. net/doc/1615099. Accessed October 2013.

Li, W.X., and M. Zhang. 2006. Effect of three-stage hypobaric storage on cell wall components, texture and cell structure of green asparagus. *Journal of Food Engineering* 77: 112–118.

Li, W.X., M. Zhang, and H. Yu. 2006a. Study on hypobaric storage of green asparagus. *Journal of Food Engineering* 77: 225–230.

Li, W.X., M. Zhang, and S.J. Wang. 2008. Effect of three-stage hypobaric storage on membrane lipid peroxidation and activities of defense enzyme in green asparagus. *LWT Food Science and Technology* 41: 2175–2181.

Liu, F.W. 1976. Storing ethylene pretreated bananas in controlled atmosphere and hypobaric air. *Journal of the American Society for Horticultural Science* 101: 198–201.

Lougheed, E.C., E.W. Franklin, D.J. Papple, D.R. Pattie, H.K. Malinowski, and A. Wenneker. 1974. *A feasibility study of low-pressure storage*. University of Guelph, Horticultural Science Department and School of Engineering, Ontario, Canada.

Lougheed, E.C., D.P. Murr, and L. Berard. 1977. LPS—great expectations. In: *Proceedings of the 2nd national controlled atmosphere research conference*, ed. D.H. Dewey, 5–7 April 1977, Michigan State University, East Lansing, pp. 38-44.

Lougheed, E.C., D.P. Murr, and L. Berard. 1978. Low pressure storage for horticultural crops. *HortScience* 13: 21–27.

Mapson, L.W., and W.G. Burton. 1962. The terminal oxidases of the potato tuber. *Biochemistry Journal* 82: 19–25.

Mbata, G.N., and T.W. Philips. 2001. Effects of temperature and exposure time on mortality of stored-product insects exposed to low pressure. *Journal of Economic Entomology* 94: 1302–1307.

McKeown, A.W., and E.C. Lougheed. 1981. Low pressure storage of some vegetables. *Acta Horticulturae* 116: 83–96.

Min, K., and C. Oogaki. 1986. Characteristics of respiration and ethylene production in fruits transferred from LP storage to ambient atmosphere. *Journal of the Japanese Society for Horticultural Science* 55: 339–347.

Mitcham, E.J., T. Martin, and S. Zhou. 2006. The mode of action of insecticidal controlled atmospheres. *Bulletin of Entomological Research* 96: 213–222.

Mori, M., and N. Kozukue. 1995. The glyalkaloid contents of potato tubers: differences in tuber greening in various varieties and lines. *Report of the Kyushu Branch of the Crop Science Society of Japan* 61: 77–79.

Navarro, S. 1978. The effects of low oxygen tensions on three stored-product insect pests. *Phytoparasitica* 6: 51–58.

Navarro, S., and M. Calderon. 1979. Mode of action of low atmospheric pressures on *Ephestia cautella* (Wlk.) pupae. *Experientia* 35: 620–621.

Navarro, S., J.E. Donahaye, R. Dias, A. Azrieli, M. Rindner, T. Phillips, R. Noyes P. Villers, T. Debruin, R. Truby, and R. Rodriguez. 2001. Application of vacuum in a transportable system for insect control. In: *Proceedings of the international conference on controlled atmosphere and fumigation in stored products,* eds. E.J. Donahaye, S. Navarro, and J.G. Leesch, Fresno, CA. 29 Oct.–3 Nov. 2000, Executive Printing Services, Clovis, CA, USA. pp. 307–315.

Navarro, S., S. Finkelman, J.E. Donahaye, A. Isikber, M. Rindner, and R. Dias. 2007. Development of a methyl bromide alternative for the control of stored product insects using a vacuum technology. In: *Proceedings of the international conference controlled atmosphere and fumigation in stored products,* eds. E.J. Donahaye, S. Navarro, C. Bell, D. Jayas, R. Noyes, and T.W. Phillips, Gold-Coast Australia. 8–13th August 2004. FTIC Ltd. Publishing, Israel, pp. 227–234.

Nilsen, K.N., and C.F. Hodges. 1983. Hypobaric control of ethylene-induced leaf senescence in intact plants of *Phaseolus vulgaris* L. *Plant Physiology* 71: 96–101.

Onoda, A., T. Koizumi, K. Yamamoto, T. Furruya, H. Yamakawa, and K. Ogawa. 1989. A study of variable low pressure storage of cabbage and turnip. *Nippon Shokuhin Kogyo Gakkaishi* 36: 369–374.

Pantastico, E.B. 1975. (ed.) *Postharvest physiology, handling and utilisation of tropical and sub-tropical fruits and vegetables*. AVI Publishing Co., Westpoint.

Patterson, M.E., and S.W. Melsted. 1977. Sweet cherry handling and storage alternatives. In: *Proceedings of the 2nd national controlled atmosphere research conference*. Michigan State University, East Lansing, p. 9.

Patterson, M.E., and S.W. Melsted. 1978. Improvement of prune quality and condition by hypobaric storage. *HortScience* 13: 351.

Pattie, D.R., and E.C. Lougheed. 1974. *Feasibility study of low pressure storage*. Department of Horticultural Science School of Engineering, University of Guelph Ontario, Canada.

Paul, A.-L., and R.J. Ferl. 2006. The biology of low atmospheric pressure—implications for exploration mission design and advanced life support. *Gravitational and Space Biology* 19: 3–17.

Prusky, D., N.T. Keen, and I. Eaks. 1983. Further evidence for the involvement of a preformed antifungal compound in the latency of *Colletotrichum gloeosporioides* on unripe avocado fruits. *Physiological Plant Pathology* 22: 189–198.

Prusky, D., H.D. Ohr, N. Grech, S. Campbell, I. Kobiler, G. Zauberman, and Y. Fuchs. 1995. Evaluation of antioxidant butylated hydroxyanisole and fungicide prochloraz for control of post-harvest anthracnose of avocado fruit during storage. *Plant Disease* 79: 797–800.

Quazi, M.H., and H.T. Freebairn. 1970. The influence of ethylene oxygen and carbon dioxide on ripening of bananas. *Botanical Gazette* 131: 5–14.

Romanazzi, G., F. Nigro, and A. Ippolito. 2003. Short hypobaric treatments potentiate the effect of chitosan in reducing storage decay of sweet cherries. *Postharvest Biology and Technology* 7: 73–80.

Romanazzi, G., F. Nigro, and Ippolito, A. 2008. Effectiveness of a short hyperbaric treatment to control postharvest decay of sweet cherries and table grapes. *Postharvest Biology and Technology* 49: 440–442.

Romanazzi, G., F. Nigro, A. Ippolito, and M. Salerno. 2001. Effect of short hypobaric treatments on postharvest rots of sweet cherries, strawberries and table grapes. *Postharvest Biology and Technology* 22: 1–6.

Ryall, A.L., and W.T. Pentzer. 1974. *Handling transportation and storage of fruits and vegetables*, vol. 2. Westport: AVI Publishing Company Incorporated.

Salunkhe, D.K., and M.T. Wu. 1975. Subatmospheric storage of fruits and vegetables. In *Postharvest biology and handling of fruits and vegetables*, ed. N.F. Haard, and D.K. Salunkhe, 153–171. Westpoint: A.V.I. Publishing Company Inc.

Salunkhe, D.K., and M.T. Wu. 1973. Effects of subatmospheric pressure storage on ripening and associated chemical changes of certain deciduous fruits. *Journal of the American Society for Horticultural Science* 98: 113–116.

SeaLand. 1991. *Shipping guide to perishables*. SeaLand Services Inc., P.O. Box 800, Iselim, New Jersey 08830, USA.

Seymour, G.B., P. John, and A.K. Thompson. 1987. Inhibition of degreening in the peel of bananas ripened at tropical temperatures. 2. Role of ethylene, oxygen and carbon dioxide. *Annals of Applied Biology* 110: 153–161.

Sharp, A.K. 1985. Temperature uniformity in a low-pressure freight container utilizing glycol-chilled walls. *International Journal of Refrigeration* 8: 37–42.

Sharples, R.O. 1971. Storage under reduced atmospheric pressure. *Annual Report of the East Malling Research Station, UK.* for 1970.

Sharples, R.O. 1974. Hypobaric storage: apples and soft fruit. *Annual Report of the East Malling Research Station, UK.* for 1973.

Sharples, R.O. and I.W. Langridge. 1973. Reduced pressure storage. *Annual Report of the East Malling Research Station, UK. for 1972* 108.

Spalding, D.H. 1980. Low pressure hypobaric storage of several fruits and vegetables. *Proceedings of the Florida State Horticultural Society* 92: 201–203.

Spalding, D.H., and W.F. Reeder. 1972. Quality of 'Booth 8' and 'Lula; avocados stored in a controlled atmosphere. *Proceedings of the Florida State Horticultural Society* 85: 337–341.

Spalding, D.H., and W.F. Reeder. 1974. Current status of controlled atmosphere storage of four tropical fruits. *Proceedings of the Florida State Horticultural Society* 87: 334–339.

Spalding, D.H., and W.F. Reeder. 1975. Low-oxygen, high carbon dioxide controlled atmosphere storage for the control of anthracnose and chilling injury of avocados. *Phytopathology* 65: 458–460.

Spalding, D.H., and W.F. Reeder. 1976. Low pressure (hypobaric) storage of limes. *Journal of the American Society for Horticultural Science* 101: 367–370.

Spalding, D.H., and W.F. Reeder. 1977. Low pressure (hypobaric) storage of mangos. *Journal of the American Society for Horticultural Science* 102: 367–369.

Staby, G.L. 1976. Hypobaric storage: an overview. *Combined Proceedings of the International Plant Propagation Society* 26: 211–215.

Staby, G.L., M.S. Cunningham, C.L. Holstead, J.W. Kelly, P.S. Konjoian, B.A. Eisenbergi, and B.S. Dressier. 1984. Storage of rose and carnation flowers. *Journal of the American Society for Horticultural Science* 109: 193–197.

Stoddard, E.S. and C.E. Hummel. 1957. Methods of improving food preservation in home refrigerators. *Refrigeration Engineering* 65:33–38, 69, 71.

Tolle, W.E. 1969. Hypobaric storage of mature green tomatoes. *US Department of Agriculture Marketing Research Report* 842.

Thompson, A.K. 2010. *Controlled atmosphere storage of fruits and vegetables*, 2nd ed. Oxford, UK: CAB International.

Thompson, A.K. 2015. *Fruit and vegetables—harvesting, handling and storage*, 3rd ed. Oxford, UK: Wiley Blackwell.

Veierskov, B., and H.G. Kirk. 1986. Senescence in oat leaf segments under hypobaric conditions. *Physiologia Plantarum* 66: 283–287.

Verlent, I., A.V. Loey, C. Smout, T. Duvetter, B.L. Nguyen, and M.E. Hendrickx. 2004. Changes in purified tomato pectinmethylesterase activity during thermal and high pressure treatment. *Journal of the Science of Food and Agriculture* 84: 1839–1847.

Wade, N.L. 1974. Effects of O_2 concentration and ethephon upon the respiration and ripening of banana fruits. *Journal of Experimental Botany* 25: 955–964.

Wang, C.S., L.S. Liang, and G.X. Wang. 2007. Effects of low pressure on quality of DongZao jujube fruit in hypobaric storage. *Food Science* 28: 335–339.

Wang, S., J. Tang, and F. Younce. 2003. Temperature measurement. In *Encyclopedia of agricultural, food, and biological engineering*, ed. D.R. Heldman, 987–993. New York: Marcel Dekker.

Wang, Z., and D.R. Dilley. 2000. Hypobaric storage removes scald-related volatiles during the low temperature induction of superficial scald of apples. *Postharvest Biology and Technology* 18: 191–199.

Ward, C.M. 1975. Hypobaric storage. *Annual Report of the National Vegetable Research Station, Wellesbourne, UK for 1974*, p. 85.

Wells, J.M. 1974. Growth of *Erwinia carotovora, E. atroseptica,* and *Pseudomonas fluorescence* in low-oxygen and high-carbon dioxide atmospheres. *Phytopathology* 64: 1012–1015.

Li, Wenxiang, Min Zhang, and Han-qing Yu. 2006b. Study on hypobaric storage of green asparagus. *Journal of Food Engineering* 73: 225–230.

Wills, R.B.H., and S.I.H. Tirmazi. 1979. Effects of calcium and other minerals on the ripening of tomatoes. *Australian Journal of Plant Pathology* 6: 221–227.

Wills, R.B.H., and S.I.H. Tirmazi. 1981. Retardation of ripening of mangoes by postharvest application of calcium. *Tropical Agriculture Trinidad* 58: 137–141.

Wills, R.B.H., and S.I.H. Tirmazi. 1982. Inhibition of ripening of avocados with calcium. *Scientia Horticulturae* 16: 323–330.

Workman, M., H.K. Pratt, and L.L. Morris. 1957. Studies on the physiology of tomato fruit. I. Respiration and ripening behaviour at 20 °C as related to date of harvest. *Proceeding of the American Society for Horticultural Science* 69: 352–365.

Wu, M.T., and D.K. Salunkhe. 1972. Subatmospheric pressure storage of fruits and vegetables. *Utah Science* 33: 29–31.

Wu, M.T., S.J. Jadhav, and D.K. Salunkhe. 1972. Effects of sub-atmospheric pressure storage on ripening of tomato fruits. *Journal of Food Science* 37: 952–956.

Yahia, E.M. 2011. Avocado (*Persea americana* Mill.). In *Postharvest biology and technology of tropical and subtropical fruits*, vol. 2, ed. E.M. Yahia, 125–185. Oxford: Woodhead Publishing.

Yuen, C.M.C. 1993. Calcium and postharvest storage potential. Postharvest handling of tropical fruit. *Australian Centre for International Agricultural Research Proceedings* 50: 218–227.

Zhao, Z., W. Jiang, J. Cao, Y. Zhao, and Y. Gu. 2006. Effect of cold-shock treatment on chilling injury in mango (*Mangifera indica* L. cv. Wacheng) fruit. *Journal of the Science of Food and Agriculture* 86: 2458–2462.

Chapter 4
Hyperbaric Storage

Introduction

As was discussed in the previous chapter, hypobaric systems have reduced partial pressure of O_2 therefore hyperbaric systems have increased partial pressure of O_2. Hyperbaric conditions are applied to fruits and vegetables and processed foods, but they are perhaps better known for their application in medicine. Hyperbaric O_2 therapy is the medical use of O_2 at levels higher than 21 kPa. The equipment usually consists of a pressure chamber with a means of delivering, usually, 100 % O_2. Their uses have included the treatment of decompression sickness, gas gangrene and carbon monoxide poisoning and they have been tested on cerebral palsy and multiple sclerosis (Mathieu 2006).

Short-time exposures to very high pressures have been used in food processing for many years. The effects of hyperbaric conditions on preserving food was described by Saraiva (2014) who commented that food that had been recovered from a sunken submarine in the 1990s was still in a consumable condition. The submarine had been sunk 10 months earlier to a depth of 1,540 m where the pressure was about 15 MPa and the temperature about 3–4 °C. Naik et al. (2013) reported that high-pressure processing was first proposed by Royer in 1895 to kill bacteria. Hite et al. (1914) reported that exposure of fruit or vegetables to 680 MPa for 10 min at room temperature gave a 5–6 log-cycle reduction in microorganism. High-pressure processing was first used commercially in the early 1990s in Japan for acid foods to be stored in chilled conditions. For fresh horticultural produce, Goyette et al. (2011) defined hyperbaric storage as "exposing fruit or vegetable to compressed air in a range lower than 10 atmospheres." Ahmed and Ramaswamy (2006) and Baba and Ikeda (2003) also defined hyperbaric treatment of fresh fruit and vegetables and stated that it is different from high-pressure treatment used in processing foods, where pressures of between 400 and 1,200 MPa are used. An application of such high pressures is generally not

© The Author(s) 2016

A.K. Thompson, *Fruit and Vegetable Storage*, SpringerBriefs in Food,
Health, and Nutrition, DOI 10.1007/978-3-319-23591-2_4

suitable for fresh fruit and vegetables because they can cause irreversible damage to cell structure. High-pressure inactivation of yeast and moulds has also been reported, for example in citrus juices that had been exposed to 400 MPa for 10 min at 40 °C. Juice treated in this way did not spoil during storage for up to 3 months (Olsson 1995). Hyperbaric conditions have also been reported to be used successfully in the storage of meat and eggs at room temperature for several days by Bert (1878 quoted by Kader and Ben-Yehoshua 2000) where compressed air at 15–44 atmospheres was used. It was reported that bacteria, yeasts and moulds were killed in various foods by this treatment. Charm et al. (1977) reported that at high pressure, the storage life of Atlantic cod was greatly extended compared to storage at atmospheric pressure. After 30 days, the fish held at -3 °C and 238 atmospheres had significantly lower bacterial counts and a higher sensory evaluation than those stored -25 °C under atmospheric pressure. Moreira et al. (2015) compared storage of soup at different pressures and temperatures. They concluded that 100 MPa compared to 150 MPa and 4 h exposure compared to 8 h resulted in more pronounced microbial growth inhibition and microbial inactivation. Aerobic mesophiles showed less susceptibility to hyperbaric storage compared to Enterobacteriaceae and yeast and moluds. Hyperbaric storage at 25 or 30 °C generally maintained the physicochemical parameters at values similar to refrigeration at 4 °C under atmospheric pressures. High pressure was reported to have potential for treatment to control quarantine insect pests in fresh or minimal processed fruits and vegetables (Butz et al. 2004).

Hyperbaric conditions, even at variable room temperatures of up to 37 °C, have been shown to preserve foods and thus achieve significant energy savings (Fernandes et al. 2015). Hyperbaric storage, at room temperature, could be more energy efficient that refrigeration since the only energy costs are during compression and no additional energy is required to subsequently maintain the product under pressure. Liplap et al. (2012) showed that hyperbaric storage of avocado fruit in ambient conditions used some 3 % of the energy required for commercial refrigerated storage at 5 °C but could have similar effects on delaying ripening. However, the capital costs of high-pressure equipment are high (Saraiva 2014), but expansion of high-pressure food processing should help to reduce equipment costs (Balasubramaniam et al. 2008).

Effects

As has been indicated above, the effects of high-pressure processing are mainly to control microorganisms but has little or no detrimental effects on flavour (Hill 1997). However, Hill quoted work where some fruit juices were affected by high-pressure processing, for example in grapefruit juice, where many consumers found the juice more acceptable after high-pressure processing because it was less bitter, but had good retention of vitamin C. Small molecules, which are the characteristics of flavouring and nutritional components, typically remained unchanged when fruit and vegetables are exposed to high pressure (Horie et al. 1991). Ludikhuyze et al. (2002) and Yordanov and Angelova (2010) reviewed the effects of high pressures on fruit and vegetable processing and reported that vegetative cells were inactivated at about 300 MPa at ambient temperature, while spore inactivation required 600 MPa or more in combination with a temperature of 60–70 °C. Previously, Larson et al. (1918 quoted by Ludikhuyze et al. 2002) had observed that pressure treatments up to 1,800 MPa at room temperature were not sufficient to affect commercial sterility of food products. A combination of pressure with temperatures of 60 °C and higher was required for extensive inactivation of spores, with the lower the pressure applied, the higher the temperature required to induce inactivation (Sale et al. 1970). At temperatures below 60 °C in combination with a pressure of about 400 MPa, there was a maximal three log-cycle reductions of spores of *Bacillus coagulans* (Roberts and Hoover 1996) and *Clostridium sporogenes* (Mills et al. 1998). Generally, Gram-positive bacteria are more resistant to pressure than Gram-negative bacteria; while fungi and yeasts and *Enterococcus hirae* was more resistant to high-pressure treatment than *Listeria monocytogenes* and *Aeromonas hydrophila*. The most resistant are bacterial spores (Fonberg-Broczek et al. 2005).

In addition to its effect on bacteria, high-pressure treatment was noted to help preserve colour in fruit and vegetable products including tomato juice (Poretta et al. 1995), orange juice (Donsi et al. 1996) guava purée (Yen and Lin 1996) and avocado purée (Lopez-Malo et al. 1998). The decrease in vitamin C content in strawberry purée and guava purée during storage after pressure treatment of 400–600 MPa for 15–30 min was much lower than loss of vitamin C content in non-pressure treated purée (Sancho et al. 1999). Luscher et al. (2005) used 250 MPa during the freeze processing at −27 °C on potatoes (*Solanum tuberosum*) and found that after thawing there was a considerable improvement in terms of texture, colour and visual appearance compared to freezing at atmospheric pressure. Goyette et al. (2007) stated that it may be possible to use much lower hyperbaric pressures for fruit and vegetables than for processed products and pressures greater than 100 MPa may be above the threshold for irreversible tissue damage, thus causing substantial injuries (Goyette et al. 2012).

There are various effects of high O_2 storage on the chemical changes and enzyme activity of some fruit and vegetables, which would be expected to be reproduced by hyperbaric conditions. The effects of hyperbaric storage appear to be mainly on the metabolism of the fruit or vegetables and on microorganisms that infect them. Under hyperbaric pressures, a large change in the respiration rate was observed immediately after the pressure was applied and its amplitude decreased during the initial period of the hyperbaric treatment, which was described as an unsteady or transient state (Goyette et al. 2012). Eggleston and Tanner (2005) found that at pressure of 600 MPa the respiration rate of carrot sticks decreased and this effect was greater the longer they were exposed over periods of 2–10 min. Liplap et al. (2014b) described the effects of hyperbaric conditions on the respiration rate of lettuce and Liplap et al. (2013a) on corn and avocados (Liplap et al. 2012). From these studies, it can be concluded that the method used to measure respiration rate was limited as it only allowed the respiration rate to be determined after the system reached equilibrium, which may require several hours. It was also proposed that gas dilution, solubilisation and/or desolubilisation processes could take place under hyperbaric pressure treatment. It is these processes that may be responsible for these apparent large changes in the respiration rate of the produce exposed to hyperbaric conditions. Liplap et al. (2013c) found that the trend in antioxidant activity observed from both O_2 radical absorbance capacity and trolox equivalent antioxidant capacity assays was generally similar. In storage of tomatoes at 13 and 20 °C there was no significant effect observed of hyperbaric exposure on lipophilic antioxidant and hydrophilic antioxidant compared with tomatoes under atmospheric pressure.

Liplap et al. (2014a, b, c) showed that bacterial growth was affected by hyperbaric pressure at 20 °C under 100, 200, 400, 625 and 850 kPa. As hyperbaric pressure increased, the bacterial growth significantly decreased, but the effect varied with species of bacteria. The maximum growth at 850 kPa was reduced by 56 % for *Pseudomonas cichorii*, by 71 % for *P. marginalis* and by 43 % for *Pectobacterium carotovorum*. In a review, San Martín et al. (2002) reported inhibitory effects of hyperbaric pressure on the growth of several microorganisms but very high pressure were required to kill them or inactivate their growth. Liplap et al. (2014b) exposed lettuce at 20 °C to pressures ranging from 100 to 850 kPa for 5 days and found that the development of decay was delayed under hyperbaric pressure, especially at 850 kPa, in comparison with those under normal atmospheric pressure. They considered the mode of action of hyperbaric pressure on the growth of microorganisms could be due to the direct impact of elevated pressure itself on the microorganisms as had been demonstrated on infections in strawberry juice (Segovia-Bravo et al. 2012). Another explanation could be that the elevated O_2 caused toxicity to bacteria, yeasts and moulds or the enhancement of defence compound synthesised by the host pathogen induced by mild stress. This latter effect of stress has been demonstrated for on tomatoes by Lu et al. (2010). High CO_2 can also affect bacterial growth and Enfors and Molin (1980) found that the growth *Bacillus cereus* was completely inhibited at three atmospheres of CO_2 and *Streptococcus cremoris* at 11 atmospheres of CO_2. Vigneault et al. (2012) commented that exposure to hyperbaric pressures could be an alternative to chemical treatment for preserving postharvest quality of fruit and vegetables.

In addition to the control of microorganisms and some reductions in susceptibility to pathogens (Baba et al. 1999; Romanazzi et al. 2008), the apparent effects of hyperbaric conditions on fresh fruit and vegetables, as well as decreased respiration rate, include decreased ethylene production, slowing of the ripening processes and the possible extension of the synthesis of certain chemicals (Baba and Ikeda 2003; Eggleston and Tanner 2005; Goyette et al. 2012). Other effects of hyperbaric exposure described by Baba and Ikeda (2003) and Goyette et al. (2012) include reduced weight loss, maintenance of peel colour and TSS:TA ratio, improved lycopene synthesis as well as some evidence of reduction in chilling injury in tomatoes. There was some indication that hyperbaric storage could improve the retention of flavour in stored fruit compared to storage under atmospheric pressure. Dong Sik Yang et al. (2009) found that the composition of volatile compounds produced by peaches after storage was higher after storage under hyperbaric pressure compared to controlled atmosphere storage or storage in air at atmospheric pressure.

Hendrickx et al. (1998) reported that exposure to increased pressures of 100 MPa or higher can induce structural rearrangements in enzymes which can cause their activation or their partial or total inactivation in a reversible or irreversible manner. The specific effect of pressure depends on several factors including the structure of the enzyme, its origin, the medium composition, pH or the temperature and pressure levels applied. For example, PME from peppers, tomatoes, white grapefruit, plums and carrots has been shown to be resistant to exposure to high pressure since pressure higher than 700 MPa is usually required to induce short-term inactivation at room temperature (Segovia-Bravo et al. 2012). In tomatoes, PG is much more pressure-labile than PME and almost complete PG inactivation was shown to occur in cherry tomatoes at 500 MPa in ambient temperature (Tangwongchai et al. 2000). Verlent et al. (2004) found that the optimal temperature for tomato pectin methylesterase activity at atmospheric pressure was about 45 °C at pH 8.0 and about 35 °C at pH 4.4, but at both pH values the optimal temperature increased as pressure was increased over the range of 0.1–600 MPa. Also at both pH values, the catalytic activity of tomato pectin methylesterase was higher at elevated pressure than at atmospheric pressure.

High Oxygen

One effect of hyperbaric conditions is to increase the partial pressure of O_2 above that found in the air at normal atmospheric pressure. For the sake of simplicity O_2 partial pressure is referred to as % O_2, where 1 % O_2 is approximately 1 kPa O_2. Some studies have been made on the effects of ways of other than hyperbaric pressures of increasing the O_2 content around fruit and vegetables. For example Day (1996), and many others, used nitrogen and O_2 premixed in the required proportions in a pressurised cylinder. This mixture was then flushed into plastic film bags at atmospheric pressure (to replace the air) with the fresh fruit or vegetable sealed within. During subsequent storage, the O_2 level within the package will then fall progressively as storage proceeds due to the respiration of the fruit or vegetable contained in the bag and any gas exchange through the plastic film. Other workers have used the same system of premixing the gases in steel cylinders and releasing them at a constant rate through a chamber in order to retain the required levels of O_2 and N_2. It has been shown by many workers that high O_2 levels in storage can affect various postharvest processes in whole fresh fruit and vegetables as well as those that have been minimally processed to be ready to eat. As well as oxidative processes, the effects of the increased O_2 levels include: respiration rate, ethylene production, volatile compounds, chlorophyll degradation, softening, pigments, nutrient content, sprouting, free radicals, diseases and physiological disorders.

Oxidation

Day et al. (1998) reported that high O_2-modified atmosphere packaging had beneficial effects on the degree of lipid oxidation. Shredded lettuce in modified atmosphere packaging stored at 5 °C for 10 days showed browning but those in 80 % O_2 + 20 % CO_2 did not (Heimdal et al. 1995). Sliced apples stored at 1 °C for 2 weeks had less browning in 100 % O_2 than in air (Lu and Toivonen (2000). Jiang (2013) coated button mushrooms (*Agaricus bisporus*) with 2 % alginate and stored them in jars continuously ventilated with 100 % O_2 at 4 °C up to 16 days and found that this treatment delayed browning. Since the browning reaction is oxidative, then it is counter intuitive that levels should be lower in high O_2. Day (1996) reported that there were highly positive effects of storing minimally processed fruit and vegetables in 70 and 80 % O_2 in the film bags. He indicated that it inhibited undesirable fermentation reactions, delayed enzymic browning and the O_2 levels of over 65 % inhibited both aerobic and anaerobic microbial growth. He also showed that the cut surface browning of apple slices were inhibited during storage in 100 % O_2 compared to those that had been stored in air. He explained this effect by suggesting that the high O_2 levels may cause substrate inhibition of PPO or alternatively, high levels of colourless quinones formed may cause feedback inhibition of PPO.

Respiration Rate

Kidd and West (1934) found that storage of apples in 100 % O_2 accelerated the onset of the climacteric rise in respiration rate and Biale and Young (1947) reported a similar increase in respiration rate in lemons exposed to 34.1, 67.5 or 99.2 % O_2. With avocados, Biale (1946) found that there was only a small acceleration in the time of the onset of the climacteric rise in respiration rate when they were exposed to 50 or 100 % O_2. Exposure of potato tubers for some hours to 100 % O_2 had little or no effect on their respiration rate, but prolonged exposure led, at first, to an increase in their respiration rate, over a period of some 2 weeks, but thereafter the effects of 'O_2 poisoning' became apparent and after 5–6 weeks the effects on respiration rate was negligible (Barker and Mapson 1955). Cherries or apricots exposed to 30, 50, 75 and 100 % O_2 showed no effect on their respiration rate but in plums the respiration rate was stimulated in proportion to the increasing O_2 concentration (Claypool and Allen 1951). However, Escalona et al. (Escalona et al. 2010) concluded that 80 % O_2 must be used in modified atmosphere packaging of fresh cut lettuce in combination with 10–20 % CO_2 to reduce their respiration rate and avoid fermentation. Zheng et al. (2008) found that Zucchini squash exposed to 100 % O_2 had the lowest respiration rate compared to storage in air or 60 % O_2.

Ethylene

Effects on increased O_2 levels during storage have been reported to have various effects on ethylene biosynthesis. Russet Burbank potatoes stored at 7 °C had a higher ethylene production rate in 80 % O_2 + 12 % CO_2 compared to storage in air (Creech et al. 1973) and Bartlett pears in storage at 20 °C had a higher ethylene production rate in 100 % O_2 than in air (Frenkel 1975). Similar results were reported by Morris and Kader (1977) for mature-green and breaker tomatoes stored at 20 °C when they were exposed to 30 or 50 % O_2 but exposure to 80 or 100 % O_2 reduced ethylene production rates and muskmelons stored in 100 % O_2 at 20 °C had similar ethylene production level as those stored in air (Altman and Corey 1987). Zheng et al. (2008) found that zucchini squash exposed to 60 % O_2 had the lowest ethylene production compared to those stored in 100 % O_2 or in air.

Volatile Compounds

It appears that exposure to high O_2 does not affect the volatile content of fruit. For example, Rosenfeld et al. (1999) found that blueberries in modified atmosphere packages stored at 4 or 12 °C for up to 17 days had similar sensory quality whether flushed with air or 40 % O_2. Yahia (1989) found that apples that had been stored at 3.3 °C in 3 % O_2 + 3 % CO_2 for up to 9 months did not have increased volatile formation when subsequently exposed to 100 % O_2 at 3.3 °C for up to 4 weeks.

Chlorophyll

Bartlett pears stored at 20 °C in 100 % O_2 had higher rates of chlorophyll degradation than those stored in air (Frenkel 1975).

Texture

Jiang (2013) coated button mushrooms with 2 % alginate and stored them in jars continuously ventilated with 100 % O_2 at 4 °C for up to 16 days and found that this treatment maintained their firmness and delayed cap opening. Additionally, the treatments delayed changes in the total soluble solids, total sugars and ascorbic acid and inhibited the activity of PPO and POD throughout storage. Bartlett pears kept at 20 °C in 100 % O_2 had higher rates of softening than those kept in air

(Frenkel 1975). Day (1996) reported that softening of slices of apples was inhibited during storage in 100 % O_2 compared to those stored in air.

Peel Spotting

Brown spots on the skins of bananas is a normal stage of ripen that usually occurs when the skin has turned from green to yellow or fully yellow depending on the variety. Maneenuam et al. (2007) compared the effect of storage in different O_2 partial pressures at 25 °C and 90 % r.h. on peel spotting in the variety Sucrier (*Musa* AA). The fruit had first been initiated to ripen and were turning yellow. They were then transferred to atmospheres containing either 90 % O_2 or 18 % O_2 in gas-tight chambers. Peel spotting and a decrease in dopamine levels (a free phenolic compound) were quicker in fruit in 90 % O_2 indicating that dopamine might be a substrate for the browning reaction. The browning reaction is related to the activity of PAL that converts phenylalanine to free phenolic substances that form the substrate that is converted to quinones by PPO.

Pigments

Biale and Young (1947) reported exposure of lemon to 99.2 % O_2 accelerated degreening and Hamlin oranges stored in 50 % O_2 had an increased rate of degreening (Jahn et al. 1969). Aharoni and Houck (1980) exposed oranges for 4 weeks at 15 °C to 40 or 80 % O_2, followed by 2 weeks in air. They found that fruits kept in 80 % O_2 had the palest coloured peel, but their endocarp and juice were the deepest orange. The response was intermediate for oranges kept in 40 %. Aharoni and Houck (1982) reported that storage in 40 or 80 kPa O_2 increased anthocyanin synthesis of flesh and juice of blood oranges cultivars. Li et al. (1973) reported that ripening of tomatoes at 12–13 °C was accelerated in 40–50 % O_2 compared with air and Frenkel and Garrison (1976) found that lycopene synthesis in *rin* tomatoes was stimulated in storage in 60 or 100 % O_2 in the presence of 10 ml L^{-1} ethylene.

Nutrition

Barker and Mapson (1952) reported that ascorbic acid content of potato tubers kept in 100 % O_2 was lower than in those stored in air. Day et al. (1998) reported that high O_2-modified atmosphere packaging had beneficial effects on the retention of ascorbic acid and degree of lipid oxidation. They also stated that high O_2-modified atmosphere packaging of minimally processed lettuce did not decrease antioxidant levels in comparison with low O_2-modified atmosphere packaging.

Sprouting

Abdel-Rahman and Isenberg (1974) found that exposure of carrots to 40 % O_2 increased sprouting and rooting during storage at 0 °C compared to those stored in air.

Free Radicals

A free radical is an element or compound that remains unaltered during its ordinary chemical changes. Increased O_2 concentrations around and within the fruit or vegetables were shown by Fridovich (1986) to result in higher levels of free radicals that can damage plant tissues.

Chilling Injury

Zheng et al. (2008) reported that there was an indication of the low chilling injury in the zucchini squash exposed to high O_2 and that the O_2 radical absorbance capacity and total phenolic levels in their skin were both induced by cold storage and further enhanced by 60 % O_2 storage. The enhanced anti-oxidative enzyme activities and the O_2 radical absorbance capacity and phenolic levels appeared to correlate with the reduced chilling injury.

Decay

Day (1996) stated that high O_2 levels could influence both aerobic and anaerobic microbial growth of microorganisms. Amanatidou et al. (1999) found that exposure to 80–90 % O_2 generally did not inhibit microbial growth strongly, but caused a significant reduction in the growth rate of some of the microorganisms they tested including *Salmonella enteritidis*, *S. typhimurium* and *Candida guilliermondii* (a yeast used in biological control). Among the ten microbial species studied, growth of some was even stimulated by high O_2. The combined application of 80–90 % O_2 plus either 20 or 10 % CO_2 had an inhibitory effect on the growth of all the microorganisms they tested. They concluded that when high O_2 or high CO_2 were applied alone, the inhibitory effect on microbial growth was highly variable, but stronger and more consistent inhibition of microbial growth occurred when the two gases are used in combination. Wszelaki and Mitcham (1999, 2000) found that 80–100 % O_2 inhibited the in vitro growth of *B. cinerea* on strawberries. However, only 100 % O_2-inhibited growth of *B. cinerea* more than 15 % CO_2 in air and then only after exposure for 14 days. No residual effect on in vitro fungal growth was observed upon transfer to air. In in vivo studies they found that *B. cinerea* on strawberries was reduced during 15 days of storage at 5 °C in 80–100 % O_2, but there was some fermentation in the fruit. However, storage in 40 % O_2 was effective and gave good control of rotting without detrimental effects on fruit quality during storage at 5 °C for 7 days. Deng et al. (2005) also reported that high O_2 storage conditions significantly reduced fruit decay in strawberries. High O_2 has been combined with other treatments for example strawberries were treated with a solution of 1 % chitosan, packaged in modified atmosphere film packages with either 80 % O_2 or 5 % O_2, with the balance nitrogen, and then stored at 4, 8, 12 and 15 °C. The coating inhibited the growth of microorganisms at all temperatures especially when combined with high O_2 that also seemed to help in retaining their colour (Tamer and Çopur 2010).

High O_2 levels have been tested on minimally processed fruit and vegetables. Day (1996) suggested that high O_2 atmospheres could be advantageous for modified atmosphere packaging by directly inhibiting the decay causing organisms, particularly fungi on soft fruits. Gonzalez-Roncero and Day (1998) reported that 99 % O_2 alone did not prevent the growth of *Pseudomonas fragi*, *Aeromonas hydrophila*, *Yersinia enterocolitica* and *Listeria monocytogenes* but they found that growth of *P. fragi* was inhibited by 14 % and *A. hydrophila* by 15 %. The combination of 80 % O_2 + 20 % CO_2 was more effective in inhibiting growth of all the organisms tested at 8 °C than either O_2 or CO_2 alone. Amanatidou et al. (2000) found that a combination of 50 % O_2 + 30 % CO_2 prolonged the shelf-life of sliced carrots by 2–3 days longer than storage in air. In in vitro studies, Caldwell (1965) reported that in some fungi and two species of bacteria it was shown that exposure to 10 atmosphere pressure of pure O_2 completely suppressed their growth. However, unlike the tissues of higher plants, the fungi, when removed from the pressure and returned to air, recovered and began to grow apparently quite normally if the period of exposure was not too long. In these cases, a period of some days normally elapsed before the growth of the colonies in air began again. Robb (1966) exposed 103 species of fungi to 10 atmosphere pressure of O_2 for 7 days and found that 52 resumed growth after exposure. Of these 52 species, 22 resumed growth after the treatment was prolonged for 14 days and included eight *Aspergillus* spp., including *A. flavus* and *A. niger* and six *Penicillium* spp.

Physiological Disorders

Kidd and West (1934) showed that storage of apples (cultivar Bramley's Seedling) in 100 % O_2 at 4 °C resulted in disorders whose symptoms included mealy flesh and browning of skin and flesh after 4 months. Solomos et al. (1997) reported that the apple cultivars Gala and Granny Smith exposed to 100 % O_2 developed extensive peel injury compared to that which occurred under lower O_2 atmospheres especially in 1 % O_2. However, Lurie et al. (1991) found that storage of the apple cultivar Granny Smith at 0 °C in 70 % O_2 for 1 month did not accelerate the severity of the disorder caused by sunscald. Production of α-farnesene and trienol, related to development of storage scald, increased in apples stored at 0 °C in 100 % O_2 atmospheres for up to 3 months. Granny Smith apples stored in 100 % O_2 were completely 'bronzed' after 3 months and contained high ethanol concentrations (Whitaker et al. 1998). An atmosphere of 100 % O_2 potentiated the effect of 0.5 ml L^{-1} ethylene on isocoumarin formation in carrots, resulting in a fivefold increase over that found in carrots treated with ethylene in air (Lafuente et al. 1996). Super-atmospheric O_2 levels increased ethylene production and the incidence and severity of pink rib and ethylene-induced russet spotting on lettuce (Klaustermeyer and Morris 1975).

Technology

Hyperbaric tanks and chambers are made mainly for the medical and diving industries for the treatment of decompression sickness, gas gangrene and carbon monoxide poisoning. As indicate above, they have also been used to improve the postharvest life of some fruit and vegetables. For example, Robitaille and Badenhop (1981) described a completely autonomous storage system with CO_2 removal and automatic O_2 replenishment for hyperbaric storage that was successfully used to store mushrooms. Goyette (2010) described the stainless steel test chamber he used which consisted of an outer chamber and inner chamber. The purpose of the outer chamber was to resist pressures up to 15 atmospheres. It was 125 mm high by 300 mm inside diameter and closed with a bolted steel cover. The internal volume of the outer chamber was 8.836 litres. An O-ring rubber seal of 6.35 mm in diameter was placed between the cover and the chamber to ensure air tightness. Two compression fittings were fastened on the side of the chamber to connect airflow inlet and outlet using plastic tubes. The inlet of the outer chamber was connected to a cylinder of compressed air with a CO_2 concentration of 325 parts per million. The cylinder was equipped with a manometer, which regulated the pressure to the desired value. The air was passed through a humidifier to increase its humidity. An airtight connector was used to insert a T-type thermocouple inside the chamber. The inner chamber is used to reduce the volume of air surrounding the produce. By reducing the volume, the time required to reach equilibrium of CO_2 concentration inside the inner chamber was also reduced. In order to measure the respiration rate of a specific produce, the volume and shape of the inner chamber may be adjusted to the produce shape and size to maintain the void volume as small as possible. The inner chamber placed inside the outer chamber was adjusted to fit one medium-sized tomato fruit. It was built using a plastic tube, 100 mm high by 75 mm in diameter. The inner chamber had a volume of 0.442

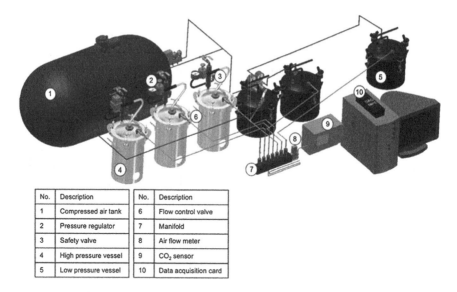

No.	Description		No.	Description
1	Compressed air tank		6	Flow control valve
2	Pressure regulator		7	Manifold
3	Safety valve		8	Air flow meter
4	High pressure vessel		9	CO_2 sensor
5	Low pressure vessel		10	Data acquisition card

Fig. 4.1 Schematic of the hyperbaric pressure respirometer system. Source: Liplap et al. (2014a, b, c) with permission

litres. The inner chamber was sealed using ABS glued fittings. An O-ring rubber seal of 3.2 mm in diameter was placed in the cover to ensure air tightness. Two compression fittings were fastened on the side of the inner chamber to connect airflow inlet and outlet. The inlet of the inner chamber was open inside of the outer chamber. A calibrated orifice inlet, 0.01 mm in diameter, was placed at the inlet of the inner chamber creating a sufficiently fine diffusion channel to avoid CO_2 dissipation from the inner chamber to the outer chamber. The outlet of the inner chamber was connected directly to the outlet of the outer chamber. Liplap et al. (2014a, b, c) describe the system they used and it is illustrated in a schematic diagram of their hyperbaric pressure respirometer system in Figs. 4.1 and 4.2.

Goyette et al. (2011) also designed a hyperbaric respirometer to explore the possibility of using hyperbaric treatment on postharvest commodities. It consisted of an airtight vessel that could be pressurised from 1 to 7 atmosphere and instrumented to automatically monitor gas concentration along time using computer controlled valves, flow meter and CO_2 gas analyser. Liplap (2014a) developed a method to determine the metabolic respiration rate of fruit or vegetables during the transient period at the beginning of a hyperbaric treatment for the correction in the apparent respiration rate by considering the dilution effect of flushing the system and the error associated with gas solubilisation as the gas partial pressure varied. They simulated the dilution process using the general equation for exhaust ventilation, thus allowing for the elimination of the dilution effect during the calculation of the net respiration rate.

Fig. 4.2 Schematic of the high-pressure vessel. Source: Liplap et al. (2014a, b, c) with permission

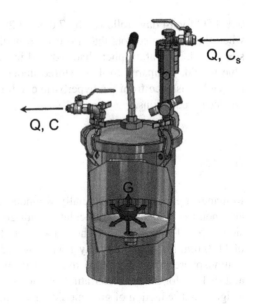

Q, C$_s$

Q, C

Horticultural Commodities

As indicated above, some work has been reported in the literature on hyperbaric treatment and storage of a few fruit and vegetables. A brief summary of some of the findings are given below.

Avocados

Liplap et al. (2012) stored avocado fruits for 7 days at ambient temperature using pressure levels of 1, 3, 5, 7 or 9 atmospheres compared to those stored under commercial cold storage conditions of 5 °C under atmospheric pressure. Hyperbaric pressure decreased ripening, resulting in an extension of the storage life and their respiration rates were inversely proportional to the pressure applied. They concluded that hyperbaric storage had potential for extending the storage life and maintaining the quality attributes of avocado while using little energy.

Cherries

Romanazzi et al. (2008) stored cherries (*Prunus avium* cultivar Ferrovia) for 24 h under pressures of 1140 mm Hg (approximately 1.5 atmospheres) and compared them with fruit stored under 1 atmosphere for 4 h. They were then stored at

0 ± 1 °C for 14 days followed by 7 days at 20 ± 1 °C under atmospheric pressure when they evaluated rots that arose from naturally occurring infections. Cherries stored under 1.5 atmosphere had reduced incidence of brown rot, grey mould and blue mould, compared to those stored under 1 atmosphere. They concluded that induced resistance from the hyperbaric conditions was likely to be responsible for the decay reductions.

Grapes

Romanazzi et al. (2008) artificially wounded grapes (*Vitis vinifera* cultivar Italia) and then the wounds were inoculated with 20 µl of a *B. cinerea* conidial suspension (5×10^4 spores ml^{-1}). The grapes were then stored for 24 h under pressures of 1140 mm Hg (approximately 1.5 atmospheres) and compared them with fruit at 1 atmosphere. They were then removed from the hypobaric chambers and stored at 2 ± 1 °C for 3 days under atmospheric pressure. Hyperbaric storage resulted in a significant reduction of grey mould lesion diameter and percentage of *B. cinerea* infections on the fruit. As with cherries, they concluded that induced resistance was likely to be responsible for the decay reductions.

Lettuce

Liplap (2013a) stored lettuces in a range of pressures from 100 to 850 kPa at 20 °C and 100 kPa at 4 °C. Hyperbaric storage at 20 °C resulted in noticeable changes in sensory quality, but they were still considered marketable after 3 days, while those stored at atmospheric pressure and 4 °C showed little degradation even during 7 days storage. After 5 days storage, the respiration rate of those under 625 and 850 kPa remained fairly stable, while the respiration rate began to increase in those in lower pressures. They concluded that this was an indication of the initiation of decay. At 20 °C, those under 850 kPa showed better quality than those under atmospheric pressure. They concluded that "overall, hyperbaric treatment has the potential of being used as an alternative technique for short-term storage of lettuce without refrigeration".

Mango

Apelbaum et al. (1977) tested the effect of hyperbaric pressure storage on mango fruit for 16 days and found that pressures from 0.25 to 0.7 MPa, (250–700 kPa) did not result in a shelf-life any greater than commercial storage in air at atmospheric pressure.

Melon Juice

Queirós et al. (2014) stored melon juice for 8 h at 25, 30 and 37 °C, under atmospheric pressure (0.1 MPa) and under hyperbaric pressures within the range of 25–150 MPa. These were compared with storage at 4 °C under atmospheric pressure. They found that hyperbaric storage at 50 and 75 MPa juice had a similar or lower microbial counts (total aerobic mesophiles, Enterobacteriaceae and yeasts/moulds) to those stored at 4 °C while at 100 and 150 MPa, the counts were lesser for all the tested temperatures, indicating an additional microbial inactivation effect. At 25 MPa no microbial inhibition was observed. Juice stored under hyperbaric conditions had similar pH, titratable acidity, total soluble solids, browning and cloudiness levels to those in storage at 4 °C.

Mume

At an average temperature of 22 °C, the shelf-life of Mume fruits (also called Japanese apricot, *Prunus mume*) was reported to be only 2 or 3 days (Miyazak 1983). Mume fruits were subjected to pressures of 0.5–5 MPa (500–5000 kPa) for 10 min and maintained at 0.5 MPa (500 kPa) for 5 days by Baba and Ikeda (2003). Treatment at 0.5 MPa decreased their respiration rate, ethylene production and weight loss during storage and showed reduced chilling injury symptoms. During storage at 0 or 5 °C, chilling injury occurred as surface pitting and/or peel browning (Goto et al. 1988). Controlled atmosphere recommendations by Koyakumaru (1997) included 25 °C with 3–5 % O_2 and 9–10 % CO_2 combined with an ethylene scrubber was effective in preserving Mume fruit (*Prunus mume*) quality during a 10 day storage period. Controlled atmosphere storage was reported by Kaji et al. (1991) to increase their storage life at 20 °C and 100 % r.h. for up to 19 days under 2–3 % O_2 with 13 % CO_2.

Mushrooms

Robitaille and Badenhop (1981) stored mushrooms under 35 atmospheres that did not affect their respiration rate, but significantly reduced moisture loss and cap browning compared to storage at normal pressure. Neither pressurisation nor gradual depressurisation over 6 h injured the mushrooms.

Peaches

The composition of volatile compounds emanating from peach fruit varied quantitatively and qualitatively during 4 weeks storage. They identified 21 compounds prior to storage and 59 after storage. Storage under hyperbaric pressure contributed most to the concentration of total volatile compounds compared to controlled atmosphere storage and storage in air (Yang et al. 2009).

Tomatoes

Romanazzi et al. (2008) observed that hyperbaric storage has been shown to have variable effects on the shelf-life of tomatoes. In their review, Ahmed and Ramaswamy (2006) reported that at 20 °C hyperbaric exposure at ≥0.3 MPa resulted in respiration rates equal or higher than those in fruit stored in ambient pressure. Liplap et al. (2013b) subjected early breaker tomatoes to ambient atmospheric pressure, 0.3, 0.5, 0.7 or 0.9 MPa at 20 °C and at 13 °C for 4 days, followed by ripening at 20 °C for up to 10 days at atmospheric pressure. Hyperbaric treatment initially inhibited lycopene synthesis but then enhanced its accumulation during exposure and subsequent ripening. All antioxidants were found in lower concentrations in tomatoes subjected to atmospheric pressure at 13 °C. They concluded that overall, hyperbaric treatment at 20 °C had potential to extend tomato shelf-life during short treatment durations without adverse impact on quality during ripening. They showed that the only consistent effect of hyperbaric treatment at 0.5, 0.7 and 0.9 MPa was to reduce weight loss and enhance firmness retention up to 5 day ripening after treatment. Hyperbaric storage at 0.5, 0.7 and 0.9 MPa significantly reduced weight loss, retained colour, firmness, total soluble solid, titratable acidity and TSS:TA ratio at similar levels to the tomato treated at 13 °C and 0.1 MPa. Firmness after treatment was highest for fruit from 0.1 MPa at 13 °C and from 0.5, 0.7 and 0.9 MPa at 20 °C. The higher firmness advantage declined during the 5th day of ripening after treatment, with higher firmness only being retained for fruit from the 0.9 MPa at 20 °C and the 0.1 MPa at 13 °C. After 10-day ripening, firmness was similar for all treatments. The lowest respiration rate was in those stored at 0.1 MPa at 13 °C. They also exposed early breaker stage tomatoes to pressures of 1, 3, 5, 7 or 9 atmospheres for 5, 10 or 15 days at 13 °C, followed by a storage at 20 °C for 12 days (Goyette et al. 2012). Based on firmness values, those at that had been stored in ambient atmospheric pressure were no longer acceptable for consumption after 12 days subsequent storage at 20 °C. Those that had been stored at 7 and 9 atmospheres for 15 days had irreversible physiological damage, while those exposed to 3, 5 or 7 atmospheres for 10 days, or 5 atmospheres for 5 days maintained marketable firmness. Lycopene content was improved in all the fruit that had been stored under hyperbaric pressures followed by 12 days of maturation compared to those that had been under atmospheric pressure with the highest lycopene content (28 % more than in those from atmospheric pressure) from fruit that had been stored under 5 atmospheres for 10 or 15 days. Goyette (2010) reported that hyperbaric treatments on tomatoes showed their respiration rate was inversely proportional to the pressure applied. Respiration rate was reduced by 20 % under 9 atmospheres compared to those under 1 atmosphere. At the onset of hypobaric storage, the respiratory quotient was low and increased to reach a value of approximately 1 within 120 h. Low respiratory quotient values were caused by solubilisation of CO_2 in the tomato cells at the beginning of the process. Liplap et al. (2013c) concluded that overall, hyperbaric treatment at 20 °C had potential to extend tomato shelf-life during short duration treatment without adverse impact on quality during ripening.

Watermelon Juice

Fidalgo et al. (2014) compared the preservation of watermelon juice at room temperature and 5 °C at atmospheric with preservation under 100 MPa at room temperature. After 8 h of hyperbaric storage at 100 MPa, the initial microbial loads of the watermelon juice were reduced by 1 log unit for total aerobic mesophiles to levels of about 3 log units and 1–2 log units for Enterobacteriaceae and yeasts and moulds that were below the detection limit. These levels remained unchanged up to 60 h. Similar results were obtained at 30 C under 100 MPa after 8 h. At atmospheric pressure for 24 h at room temperature and for 8 h at 30 °C, microbial levels were above quantification limits and unacceptable for consumption. Storage at 5 °C after the hyperbaric exposure gave an extended shelf-life.

References

Abdel-Rahman, M., and F.M.R. Isenberg. 1974. Effects of growth regulators and controlled atmosphere on stored carrots. *Journal of Agricultural Science* 82: 245–249.

Aharoni, Y., and L.G. Houck. 1980. Improvement of internal color of oranges stored in oxygen-enriched atmospheres. *Scientific Horticulture* 13: 331–338.

Aharoni, Y., and L.G. Houck. 1982. Change in rind, flesh, and juice color of blood oranges stored in air supplemented with ethylene or in O_2-enriched atmospheres. *Journal of Food Science* 47: 2091–2092.

Ahmed, J., and H.S. Ramaswamy. 2006. High pressure processing of fruits and vegetables. *Stewart Postharvest Review* 1: 1–10.

Altman, S.A., and K.A. Corey. 1987. Enhanced respiration of muskmelon fruits by pure oxygen and ethylene. *Scientific Horticulture* 31: 275–281.

Amanatidou, A., E.J. Smid, and L.G.M. Gorris. 1999. Effect of elevated oxygen and carbon dioxide on the surface growth of vegetable-associated micro-organisms. *Journal of Applied Microbiology* 86: 429–438.

Amanatidou, A., R.A. Slump, L.G.M. Gorris, and E.J. Smid. 2000. High oxygen and high carbon dioxide modified atmospheres for shelf-life extension of minimally processed carrots. *Journal of Food Science* 65: 61–66.

Apelbaum, A., and R. Barkai-Golan. 1977. Spore germination and mycelia growth of postharvest pathogens under hypobaric pressure. *Phytopathology* 67: 400–403.

Baba, T., and F. Ikeda. 2003. Use of high pressure treatment to prolong the postharvest life of mume fruit (*Prunus mume*). *Acta Horticulturae* 628: 373–377.

Baba, T., G. Como, T. Ohtsubo, and F. Ikeda. 1999. Effects of high pressure treatment on mume fruit (*Prunus mume*). *Journal of the American Society for Horticultural Science* 124: 399–401.

Baba, T., S. Ito, F. Ikeda, and M. Manago. 2008. Effectiveness of high pressure treatment at low temperature to improve postharvest life of horticultural commodities. *Acta Horticulturae* 768: 417–422.

Balasubramaniam, V.M., D. Farkas, and E.J. Turek. 2008. Preserving foods through high-pressure processing. *Food Technology* 62: 32–38.

Barker, J., L.W. Mapson. 1952. The ascorbic acid content of potato tubers. III. The influence of storage in nitrogen, air and pure oxygen. *New Phytologist* 51: 90–115.

Barker, J., and L.W. Mapson. 1955. Studies in the respiratory and carbohydrate metabolism of plant tissues. VII. Experimental studies with potato tubers of an inhibition of the respiration and of a 'block' in the tricarboxylic acid cycle induced by 'oxygen poisoning'. *Proceedings of the Royal Society Biology* 143: 523–549.

Biale, J.B. 1946. Effect of oxygen concentration on respiration of the Fuerte avocado fruit. *American Journal of Botany* 33: 363–373.

Biale, J.B. 1950. Postharvest physiology and biochemistry of fruits. *Annual Review of Plant Physiology* 1: 183–206.

Biale, J.B., and R.E. Young. 1947. Critical oxygen concentrations for the respiration of lemons. *American Journal of Botany* 34: 301–309.

Butz, P., Y. Serfert, G.A. Fernandez, S. Dieterich, R. Lindaeur, A. Bognar, et al. 2004. Influence of high pressure treatment at 25 and 80 °C on folates in orange juice and model media. *Journal of Food Science* 69: 117–121.

Caldwell, J. 1965. Effects of high partial pressures of oxygen on fungi and bacteria. *Nature* 206: 321–323.

Charm, S.E., H.E. Longmaid, and J. Carver. 1977. Simple system for extending refrigerated, nonfrozen preservation of biological-material using pressure. *Cryobiology* 14: 625–636.

Claypool, L.L., and F.W. Allen. 1951. The influence of temperature and oxygen level on the respiration and ripening of Wickson plums. *Hilgardia* 121: 29–160.

Creech, D.L., M. Workman, and M.D. Harrison. 1973. The influence of storage factors on endogenous ethylene production by potato tubers. *American Potato Journal* 50: 145–150.

Day, B.P.F. 1996. High O_2 modified atmosphere packaging for fresh prepared produce. *Postharvest News and Information* 7(31N): 34N.

Day, B.P.F., W.J. Bankier, and M.I. Gonzalez. 1998. Novel modified atmosphere packaging (MAP) for fresh prepared produce. Research Summary Sheet 13, Campden and Chorleywood Food Research Association, Chipping Campden, UK.

Deng, Y., Y. Wu, and Y. Li. 2005. Effects of high O_2 levels on post- harvets quality and shelf life of table grapes during long-term storage. *European Food Research and Technology* 221: 392–397.

Donsi, G., G. Ferrari, and M. Di Matteo. 1996. High pressure stabilization of orange juice: Evaluation of the effects of process conditions. *Italian Journal of Food Science* 2: 99–106.

Eggleston, V., and D.J. Tanner. 2005. Are carrots under pressure still alive?—The effect of high pressure processing on the respiration rate of carrots. *Acta Horticulturae* 687: 371–373.

Enfors, S.O., and G. Molin. 1980. Effect of high concentrations of carbon dioxide on growth rate of *Pseudomonas fragi, Bacillus cereus* and *Streptococcus cremoris*. *Journal of Applied Bacteriology* 48: 409–416.

Escalona, V.H., E. Aguayo, G.B. Martínez-Hernández, and F. Artés. 2010. UV-C doses to reduce pathogen and spoilage bacterial growth *in vitro* and in baby spinach. *Postharvest Biology and Technology* 56: 223–231.

Fernandes, P.A.R., S.A. Moreira, L.G. Fidalgo, M.D. Santos, R.P. Queirós, I. Delgadillo, and J.A. Saraiva. 2015. Food Preservation under pressure (hyperbaric storage) as a possible improvement/alternative to refrigeration. *Food Engineering Reviews* 7: 1–10.

Fidalgo, L.G., M.D. Santos, R.P. Queirós, R.S. Inácio, M.J. Mota, R.P. Lopes, M.S. Gonçalves, R.F. Neto, and J.A. Saraiva. 2014. Hyperbaric storage at and above room temperature of a highly perishable food. *Food and Bioprocess Technology* 7: 2028–2037.

Fonberg-Broczek, M., B. Windyga, J. Szczawinski, M. Szczawinska, D. Pietrzak, and G. Prestamo. 2005. High pressure processing for food safety. *Acta Biochimica Polonica* 52: 721–724.

Frenkel, C. 1975. Oxidative turnover of auxins in relation to the onset of ripening in Bartlett pear. *Plant Physiology* 55: 480–484.

Frenkel, C., and S.A. Garrison. 1976. Initiation of lycopene synthesis in the tomato mutant *rin* as influenced by oxygen and ethylene interactions. *HortScience* 11: 20–21.

Fridovich, I. 1986. Biological effects of the superoxide radical. *Archives Biochemistry Biophysics* 247: 1–11.

Gonzalez-Roncero, M.I., and B.P.F. Day. 1998. The effect of elevated oxygen and carbon dioxide modified atmosphere on psychotrophic pathogens and spoilage microorganisms associated with fresh prepared produce. Research Summary Sheet 98, Campden and Chorleywood Food Research Association, Chipping Campden, UK.

Goto, M., T. Minamide, and T. Iwata. 1988. The change in chilling sensitivity in fruits of mume Japanese apricot, *Prunus mume* Sieb. et Zucc. depending on maturity at harvest and its relationship to phospholipid composition and membrane permeability. *Journal of the Japanese Society for Horticultural Science* 56: 479–485.

Goyette, B., M.T. Charles, C. Vigneault, and G.S.V. Raghavan. 2007. Pressure treatment for increasing fruit and vegetable qualities. *Stewart Postharvest Review* 3: 5.1–5.6.

Goyette, B. 2010. Hyperbaric treatment to enhance quality attributes of fresh horticultural produce. PhD thesis, McGill University, Montreal, Canada.

Goyette, B., C. Vigneault, N. Wang, and G.S.V. Raghavan. 2011. Conceptualization, design and evaluation of a hyperbaric respirometer. *Journal of Food Engineering* 105: 283–288.

Goyette, B., C. Vigneault, M.T. Charles, and G.S.V. Raghavan. 2012. Effect of hyperbaric treatments on the quality attributes of tomato. *Canadian Journal of Plant Science* 92: 541–551.

Heimdal, H., B.F. Kuhn, L. Poll, and L.M. Larsen. 1995. Biochemical changes and sensory quality of shredded and MA-packaged iceberg lettuce. *Journal of Food Science* 60: 1265–1268.

Hendrickx, M., L. Ludikhuyze, I. Van den Broeck, and C. Weemaes. 1998. Effects of high pressure on enzymes related to food quality. *Trends in Food Science & Technology* 9: 197–203.

Hill, S. 1997. Squeezing the death out of food. *New Scientist* 2077: 28–32.

Hite, B.H., N.J. Giddings, and C.W. Weakley. 1914. The effect of pressure on certain microorganisms encountered in the preservation of fruits and vegetables. *Bulletin, West Virginia University Experimental Station USA* 146: 3–67.

Horie, Y., K. Kimura, M. Ida, Y. Yosida, and K. Ohki. 1991. Jam preservation by pressure pasteurization. *Nippon Nogeiku Kaichi* 65: 975–980.

Jahn, O.L., W.G. Chace, and R.H. Cubbedge. 1969. Degreening of citrus fruits in response to varying levels of oxygen and ethylene. *Journal of the American Society for Horticultural Science* 94: 123–125.

Jiang, T. 2013. Effect of alginate coating on physicochemical and sensory qualities of button mushrooms (*Agaricus bisporus*) under a high oxygen modified atmosphere. *Postharvest Biology and Technology* 76: 91–97.

Jin, A.X., Y.P. Wang, and L.S. Liang. 2006. Effects of atmospheric pressure on the respiration and softening of DongZao jujube fruit during hypobaric storage. *Journal of the Northwest Forestry University* 21: 143–146.

Kader, A.A., and S. Ben-Yehoshua. 2000. Effects of superatmospheric oxygen levels on postharvest physiology and quality of fresh fruits and vegetables. *Postharvest Biology and Technology* 20: 1–13.

Kaji, H., T. Ikebe, and Y. Osajima. 1991. Effects of environmental gases on the shelf life of Japanese apricot. *Journal of the Japanese Society for Food Science and Technology* 38: 797–803.

Kidd, F., and C. West. 1934. Injurious effects of atmospheres of pure O_2 upon apples and pears at low temperatures. *Report of the Food Investigation Board, London, UK, for 1933*, 74–77.

Klaustermeyer, J.A., and L.L. Morris. 1975. The effects of ethylene and carbon monoxide on the induction of russet spotting on crisphead lettuce. *Plant Physiology* 56(Supplement): 63.

Koyakumaru, T. 1997. Effects of temperature and ethylene removing agents on respiration of mature green mume Prunus mume Sieb. et Zucc. fruit held under air and controlled atmospheres. *Journal of the Japanese Society for Horticultural Science* 66: 409–418.

Lafuente, M.T., G. Lopez-Galvez, M. Cantwell, and S.F. Yang. 1996. Factors influencing ethylene-induced isocoumar information and increased respiration in carrots. *Journal of the American Society for Horticultural Science* 121: 537–542.

Li, Y., V.H. Wang, C.Y. Mao, and C.H. Duan. 1973. Effects of oxygen and carbon dioxide on after ripening of tomatoes. *Acta Botanica Sinica* 15: 93–102.

Liplap, P., C. Vigneault, G.S. Vijaya Raghavan, and S. Jenni. 2012. Storing avocado under hyperbaric pressure, Poster Board #035. American Society for Horticultural Science, Annual Conference August 2012.

Liplap, P., D. Charlebois, M.T. Charles, P. Toivonen, C. Vigneault, and G.S. Vijaya Raghavan. 2013a. Tomato shelf-life extension at room temperature by hyperbaric pressure treatment. *Postharvest Biology and Technology* 86: 45–52.

Liplap, P., C. Vigneault, P. Toivonen, J. Boutin, and G.S. Vijaya Raghavan. 2013b. Effect of hyperbaric treatment on respiration rates and quality attributes of sweet corn. *International Journal of Postharvest Technology and Innovation* 3: 257–271.

Liplap, P., C. Vigneault, P. Toivonen, M.T. Charles, and G.S. Vijaya Raghavan. 2013c. Effect of hyperbaric pressure and temperature on respiration rates and quality attributes of tomato. *Postharvest Biology and Technology* 86: 240–248.

Liplap, P., J. Boutin, D.I. LeBlanc, C. Vigneault, and G.S.V. Raghavan. 2014a. Effect of hyperbaric pressure and temperature on respiration rates and quality attributes of Boston lettuce. *International Journal of Food Science & Technology* 49: 137–145.

Liplap, P., P. Toivonen, C. Vigneault, J. Boutin, and G.S. Vijaya Raghavan. 2014b. Effect of hyperbaric pressure treatment on the growth and physiology of bacteria that cause decay in fruit and vegetables. *Food Bioprocess Technology* 7: 2267–2280.

Liplap, P., C. Vigneault, T.J. Rennie, J. Boutin, and G.S. Vijaya Raghavan. 2014c. Method for determining the respiration rate of horticultural produce under hyperbaric treatment. *Food Bioprocess Technology* 7: 2461–2471.

Lopez-Malo, A., E. Palou, G.V. Barbosa-Canovas, J. Welti-Chanes, and B.G. Swanson. 1998. Polyphenoloxidase activity and colour changes during storage of high hydrostatic pressure treated avocado purée. *Food Research International* 31: 549–556.

Lu, C., and P.M.A. Toivonen. 2000. Effect of 1 and 100 kPa O_2 atmospheric pretreatment of whole 'Spartan' apples on subsequent quality and shelf-life of slices stored in modified atmosphere packages. *Postharvest Biology and Technology* 18: 99–107.

Lu, J., M.T. Charles, C. Vigneault, B. Goyette, and G.S.V. Raghavan. 2010. Effect of heat treatment uniformity on tomato ripening and chilling injury. *Postharvest Biology and Technology* 56: 155–162.

Ludikhuyze, L., A. Van Loey Indrawati, and M. Hendrickx. 2002. High pressure processing of fruit and vegetables. In *Fruit and vegetable processing—improving quality,* ed. Jongen, W., 346–359. Cambridge: Woodhead Publishing Limited, Abington Hall.

Lurie, S., E. Pesis, and R. Ben-Arie. 1991. Darkening of sunscald on apples in storage is a non-enzymic and non-oxidative process. *Postharvest Biology and Technology* 1: 119–125.

Luscher, C., O. Schluter, and D. Knorr. 2005. High pressure-low temperature processing of foods: impact on cell membranes, texture, colour and visual appearance of potato tissue. *Innovative Food Science and Emerging Technologies* 6: 59–71.

Maneenuam, T., S. Ketsa, and W.G. Van Doorn. 2007. High oxygen levels promote peel spotting in banana fruit. *Postharvest Biology and Technology* 43: 128–132.

Mathieu, D. 2006. *Handbook on hyperbaric medicine*. Berlin: Springer.

Mills, G., R. Earnshaw, and M.F. Patterson. 1998. Effects of high hydrostatic pressure on *Clostridiuim sporogenes* spores. *Letters Applied Microbiology* 26: 227–230.

Miyazaki, T. 1983. Effects of seal packaging and ethylene removal from sealed bags on the shelf life of mature green Japanese apricot *Prunus mume* Sieb. Zucc. fruits. *Journal of the Japanese Society for Horticultural Science* 52: 85–92.

Moreira, S.A., P.A.R. Fernandes, R. Duarte, D.I. Santos, L.G. Fidalgo, M.D. Santos, R.P. Queirós, I. Delgadillo, and J.A. Saraiva. 2015. A first study comparing preservation of a ready-to-eat soup under pressure (hyperbaric storage) at 25 °C and 30 °C with refrigeration. *Food Science & Nutrition*. doi:10.1002/fsn3.212.

Morris, L., and A.A. Kader. 1977. Physiological disorders of certin vegetables in relation to modified atmosphere. *Second national controlled atmosphere research conference. Proceedings, Michigan State University Horticultural Report* 28, pp. 266–267.

Naik, L., R. Sharma, Y.S. Rajput, and G. Manju. 2013. Application of high pressure processing technology for dairy food preservation-future perspective: A review. *Journal of Animal Production Advances* 3: 232–241.

Olsson, S. 1995. Production equipment for commercial use. In *High pressure processing of foods*, eds. D.A. Ledward, D.E. Johnston, R.G. Earnshaw, A.P.M. Hasting, 167. Nottingham: Nottingham University Press.

Poretta, S., A. Birzi, C. Ghizzoni, and E. Vicini. 1995. Effects of ultra-high hydrostatic pressure treatments on the quality of tomato juice. *Food Chemistry* 52: 35–41.

Queirós, R.P., M.D. Santos, L.G. Fidalgo, M.J. Mota, R.P. Lopes, R.S. Inácio, I. Delgadillo, and J.A. Saraiva. 2014. Hyperbaric storage of melon juice at and above room temperature and comparison with storage at atmospheric pressure and refrigeration. *Food Chemistry* 147: 209–214.

Robb, S.M. 1966. Reactions of fungi to exposure to 10 atmospheres pressure of oxygen. *Journal of General Microbiology* 45: 17–29.

Roberts, C.M., and D.G. Hoover. 1996. Sensitivity of *Bacillus coagulans* spores to combinations of high hydrostatic pressure, heat, acidity and nisin. *Journal of Applied Bacteriology* 81: 363–368.

Robitaille, H.A., and A.F. Badenhop. 1981. Mushroom response to postharvest hyperbaric storage. *Journal of Food Science* 46: 249–253.

Romanazzi, G., F. Nigro, and A. Ippolito. 2008. Effectiveness of a short hyperbaric treatment to control postharvest decay of sweet cherries and table grapes. *Postharvest Biology and Technology* 49: 440–442.

Rosenfeld, H.J., K.R. Meberg, K. Haffner, and H.A. Sundell. 1999. MAP of highbush blueberries: sensory quality in relation to storage temperature, film type and initial high oxygen atmosphere. *Postharvest Biology and Technology* 16: 27–36.

Sale, A.J.H., G.W. Gould, and W.A. Hamilton. 1970. Inactivation of bacterial spores by hydrostatic pressure. *Journal of General Microbiology* 60: 323–334.

San Martín, M.F., G.V. Barbosa-Cánovas, and B.G. Swanson. 2002. Food processing by high hydrostatic pressure. *Critical Reviews in Food Science and Nutrition* 42: 627–645.

Sancho, F., Y. Lambert, G. Demazeau, A. Largeteau, J.-M. Bouvier, and J.-F. Narbonne. 1999. Effects of ultra-high hydrostatic pressure on hydrosoluble enzymes. *Journal of Food Engineering* 39: 247–253.

Saraiva, J.A. 2014. Storage of foods under mild pressure (hyperbaric storage) at variable (uncontrolled) room: A possible new preservation concept and an alternative to refrigeration? *Journal of Food Processing & Technology* 5: 56. doi:10.4172/2157-7110.S1.002 (Abstract).

Segovia-Bravo, K.A., B. Guignon, A. Bermejo-Prada, P.D. Sanz, and L. Otero. 2012. Hyperbaric storage at room temperature for food preservation: a study in strawberry juice. *Innovative Food Science and Emerging Technologies* 15: 14–22.

Solomos, T., B. Whitaker, and C. Lu. 1997. Deleterious effects of pure oxygen on 'Gala' and 'Granny Smith' apples. *HortScience* 32: 458.

Tamer, C.E., and O.U. Çopur. 2010. Chitosan: an edible coating for fresh-cut fruits and vegetables. *Acta Horticulturae* 877: 619–624.

Tangwongchai, R., D.A. Ledward, and J.M. Ames. 2000. Effect of high-pressure treatment on the texture of cherry tomato. *Journal of Agricultural and Food Chemistry* 48: 1434–1441.

Verlent, I., A.V. Loey, C. Smout, T. Duvetter, B.L. Nguyen, and M.E. Hendrickx. 2004. Changes in purified tomato pectinmethylesterase activity during thermal and high pressure treatment. *Journal of the Science of Food and Agriculture* 84: 1839–1847.

Vigneault, C., D.I. Leblanc, B. Goyette, and S. Jenni. 2012. Invited review: Engineering aspects of physical treatments to increase fruit and vegetable phytochemical content. *Canadian Journal of Plant Science* 92: 372–397.

Whitaker, B.D., T. Solomos, and D.J. Harrison. 1998. Synthesis and oxidation of α-farnesene during high and low O_2 storage of apple cultivars differing in scald susceptibility. *Acta Horticulturae* 464: 165–170.

Wszelaki, A.L., and E.J. Mitcham. 1999. Elevated oxygen atmospheres as a decay control alternative on strawberry. *HortScience* 34: 514–515.

Wszelaki, A.L., and E.J. Mitcham. 2000. Effects of super-atmospheric oxygen on strawberry fruit quality and decay. *Postharvest Biology and Technology* 20: 125–133.

Yahia, E.M. 1989. CA storage effect on the volatile flavor components of apples. In: *Proceedings of the fifth international controlled atmosphere research conference, Wenatchee, Washington, USA*, vol. 1, pp. 341–352, 14–16 June 1989.

Yang, Dong Sik, R.R. Balandrán-Quintana, C.F. Ruiz, R.T. Toledo, and S.J. Kays. 2009. Effect of hyperbaric, controlled atmosphere and UV treatments on peach volatiles. *Postharvest Biology and Technology* 51: 334–341.

Yen, G.-C., and H.-T. Lin. 1996. Comparison of high pressure treatment and thermal pasteurisation effects on the quality and shelf life of guava purée. *International Journal of Food Science & Technology* 31: 205–213.

Yordanov, D.G., and G.V. Angelova. 2010. High pressure processing for foods preserving. *Biotechnology and Biotechnological Equipment* 3: 1940–1945.

Zheng, Y., R.W. Fung, S.Y. Wang, and C.Y. Wang. 2008. Transcript levels of antioxidative genes and oxygen radical scavenging enzyme activities in chilled zucchini squash in response to superatmospheric oxygen. *Postharvest Biology and Technology* 47: 151–158.

Chapter 5
Conclusions

The importance of fruit and vegetables in the human diet has been known from time immemorial, but the reasons for their importance have been increasingly major areas of research leading to an understanding of the various chemicals involved in human development, health and control of diseases. Also the development in the technology of the preservation of fresh fruit and vegetables has developed and is increasingly important. This is partly because of a relentlessly increasing human population with an increasing demand for improved health and quality of life. Human beings have developed techniques over millennia for preserving fruit and vegetables, but only in the last, couple of centuries have techniques been developed based on scientific research. Control of temperature and gaseous atmosphere has been exploited and their effects better understood. Technologies have been developed and applied but these need to be constantly refined and reapplied to meet human requirements of food security and health. When controlled atmosphere storage was first introduced, the level of the various gases was predetermined by experiment and applied in practice with a view to the technically possible as well as the optimum requirements of the crop. Developments have been achieved where the metabolism of the crop can be linked to computer controls that can result in better and longer storage. Implementations of hypobaric and hyperbaric conditions in the store are ways that have perhaps been under-exploited in preservation of fresh fruit and vegetables. In theory, hypobaric exposure can have considerable benefits when used in their storage. Positive effects have been shown by many workers over a protracted period, and in one case there was considerable investment in the technology especially for international transport, but unfortunately with substantial financial losses. Effects of hypobaric exposure are similar in many ways to controlled atmosphere storage but precise controls can be simpler and there are added effects that are unique to hypobaric storage. Hypobaric exposure can also control postharvest pests and diseases in fruit and vegetables including insect pests that may be transported from country

© The Author(s) 2016

A.K. Thompson, *Fruit and Vegetable Storage*, SpringerBriefs in Food,
Health, and Nutrition, DOI 10.1007/978-3-319-23591-2_5

to country (Jiao et al. 2013) as well as some postharvest diseases (Spalding and Reeder 1976). These are effects that can possibly replace postharvest chemical treatment, which are increasingly unacceptable, and it may be possible, in the future, to use exposure to hypobaric conditions as a quarantine treatment. So the question remains why has it not been widely adopted commercially as has controlled atmosphere storage and modified atmosphere packaging?

Hyperbaric treatment and storage have been known and used on fruit and vegetables for processing for many decades, but information its use postharvest for fresh fruit and vegetables is limited. It can be effectively used in the control of disease-causing organisms infecting fruit and vegetables and, like hyperbaric conditions cited above, it has the potential as a treatment to control quarantine insect pests (Butz et al. 2004). Perhaps, the high cost of equipment and technology used for hyperbaric as well as hypobaric treatment and storage may remain the factor limiting their use in the fresh fruit and vegetable postharvest industry. Saraiva (2014) concluded that hyperbaric storage at room temperature could contribute to energy savings in the cold chain since energy costs of hyperbaric storage are only during compression. Maintaining the product under pressure does not involve any additional energy requirement. Vigneault et al. (2012) commented that exposure to hyperbaric pressures could be an alternative to chemical treatment for preserving postharvest quality of fruit and vegetables, which, as indicated above, is becoming increasingly unacceptable. Nevertheless, more research is needed before giving categorical conclusions about the potential of hyperbaric storage, especially to investigate the behavior of selected microorganisms for long periods under pressure, to study the activity of different enzymes involved in food spoilage and to evaluate both the capital and the operating costs involved.

It may be concluded that both these techniques retain potential for application to address quantitative and qualitative challenges in the postharvest sector of the fresh fruit and vegetables industry.

What is the future for fruit and vegetables storage? Over the last couple of centuries, or so, increasing awareness of the changes that occur postharvest have led to practical developments in storage. These in turn have given improved quality, nutritional content and seasons of availability that have benefitted mankind. Presumably, more sensitive control of the temperature and gaseous environment will be continually modified and improved. Also the possible replacement of these parameters with changes in the pressure within the storage environment may replace or perhaps, more likely, supplement the conditions that are currently standard practice in the fruit and vegetables industry. We live in interesting times.

References

Butz, P., Y. Serfert, G.A. Fernandez, S. Dieterich, R. Lindaeur, A. Bognar, et al. 2004. Influence of high pressure treatment at 25 and 80 °C on folates in orange juice and model media. *Journal of Food Science* 69: 117–121.

Jiao, S., J.A. Johnson, J. Tang, D.S. Mattinson, J.K. Fellman, T.L. Davenport, and S. Wang. 2013. Tolerance of codling moth and apple quality associated with low pressure/low temperature treatments. *Postharvest Biology and Technology* 85: 136–140.

Saraiva JA. Storage of foods under mild pressure (hyperbaric storage) at variable (uncontrolled) room: A possible new preservation concept and an alternative to refrigeration? *Journal of Food Science and Technology* 2014;5:56. doi: 10.4172/2157-7110.S1.002 (Abstract).

Spalding, D.H., and W.F. Reeder. 1976. Low pressure (hypobaric) storage of limes. *Journal of the American Society for Horticultural Science* 101: 367–370.

Vigneault, C., D.I. Leblanc, B. Goyette, and S. Jenni. 2012. Invited review: engineering aspects of physical treatments to increase fruit and vegetable phytochemical content. *Canadian Journal of Plant Science* 92: 372–397.

About the Author

Anthony Keith Thompson Formerly: Professor of Plant Science, University of Asmara, Eritrea; Visiting Professor of Horticulture, Hamelmalo Agriculture College, Keren, Eritrea; Professor of Postharvest Technology, Cranfield University, UK; Team Leader, European Union project at the Windward Islands Banana Development and Exporting Company; Principal Scientific Officer, Tropical Products Institute, London; Team Leader and Expert for the United Nations Food and Agriculture Organization in the Sudan and Korea; Advisor to the Colombian Government and the Jamaican Government in postharvest technology of fruit and vegetables; Research Fellow in Crop Science, University of the West Indies, Trinidad; Demonstrator in Biometrics, University of Leeds, UK.

© The Author(s) 2016
A.K. Thompson, *Fruit and Vegetable Storage*, SpringerBriefs in Food,
Health, and Nutrition, DOI 10.1007/978-3-319-23591-2

Index

© The Author(s) 2016
A.K. Thompson, *Fruit and Vegetable Storage*, SpringerBriefs in Food,
Health, and Nutrition, DOI 10.1007/978-3-319-23591-2

Printed in the United States
By Bookmasters